Susan Mufford

859-509-3191

NEW WORLDS BEYOND THE ATOM

New Worlds Beyond the Atom

BY LANGSTON DAY

IN COLLABORATION WITH

GEORGE DE LA WARR

Republished by
EP Publishing Limited
1973

Republished 1973 by EP Publishing Limited
East Ardsley, Wakefield
Yorkshire, England

Offset by permission of Robinson & Watkins Ltd. First published 1956 by Vincent Stuart Publishers Ltd.

ISBN 0 85409 893 3

Please address all enquiries to EP Publishing Ltd.
(address as above)

Reprinted in Great Britain by
The Scolar Press Limited, Menston, Yorkshire

CONTENTS

FOREWORD

When an important scientific discovery is made, articles describing it first appear in scientific and technical publications, after which books are written for the students of the new subject; and finally popular expositions become available for the lay reader.

This book represents a departure from the established practice, for it is addressed to the general public and it describes discoveries which do not yet form a part of orthodox science. Indeed the phenomena involved are so baffling and appear to be at variance with so many known facts that science will have to expand its present frontiers considerably before it is able to embrace them. But this is bound to take a long time, and so it was decided to put something into print, for it was felt that the importance of the discoveries set out herein was so great that it would be a disservice to the public to withhold from them knowledge of the facts involved and their significance.

This book is an attempt, handicapped by the limitations of present-day language, to present an accurate record of what happened between the years 1942 and 1955. Certain theories are advanced, but no claim is made for their correctness or finality: they merely indicate the trains of thought of the experimenters. But undoubtedly, as scientists probe deeper into the mysteries of the subject, further facts will be revealed and new theories will be formulated.

PART I

CHAPTER ONE

HOW MUCH DOES SCIENCE KNOW?

> It is impossible to meditate on time and the mystery of the creative passage of nature without an overwhelming emotion at the limitations of the human intelligence.
> PROFESSOR A. N. WHITEHEAD, in *Concept of Nature*

IN this book certain discoveries will be described which have been made by a team of scientists working in Oxford. These discoveries have penetrated beyond the atom to the origin of matter and are undoubtedly the beginning of a new chapter of science. To appreciate their importance it is necessary to understand the rather bewildering position in which science finds itself today.

For at least a century science has been elevated to the status of a religion. 'Genesis is not in it with a school text-book on chemistry,' says Professor J. H. Woodger. This dogmatism dates from the days of our grandfathers who imagined the Universe as a huge machine made of good, hard, indivisible atoms. Thinking was a vibration of molecules in the brain. Consciousness was an 'epiphenomenon'. Everything progressed under the impact of the mechanical forces of evolution.

Today, after some fifty years of the New Physics, most of these matter-of-fact ideas have been discredited. As Professor Whitehead says*:

> The Progress of Science has now reached a turning point. The stable foundations of Physics have broken up. . . . The old foundations of scientific thought are becoming unintelligible. Time, space, matter, material, ether, electricity, mechanism, organism, configuration, structure, pattern, function, all require reinterpretation. What is the sense of talking about a mechanical explanation when you do not know what you mean by mechanics?

* *Science and the Modern World*, by Professor A. N. Whitehead.

Our success in applied science gives us the illusion that we are gaining knowledge of Reality. But in the light of the New Physics of Einstein and others, everything which was once thought to be so firm and irrevocable is dissolving. On the plane of the sub-microscopic and on the shifting sands of relativity there is no longer certainty about the real nature of anything.

What is an atom, the force of gravity, heat, light, magnetism or electricity? One thing is explicable in terms of another thing, which can be explained only in terms of a third thing, and so on until we get back to the first thing again. Professor Eddington calls it a 'dance of relata' and he says: 'The eternal world of physics has become a world of shadows.'

According to some thinkers of the materialist school, the Universe came into existence by a fortuitous concourse of atoms. (How the atom arose they do not say.) But if this is true, how did order arise out of chaos?

A very ancient idea is that Creation is based on a Law of Harmonics, one aspect of which is seen clearly in music. Bode's Law seems to bear this out as far as the planets are concerned, while the table of atomic weights confirms the fact that it applies to matter. Atoms, according to the current view, are analogous to planetary systems. Organic Life is a single entity, beautifully designed and balanced, with a close-working system of reciprocal maintenance. The galaxies have definite structure, and so no doubt has the Universe as a whole. Creation is an orderly process which implies the existence of laws. But what are these laws?

The New Physics attaches great importance to 'Action', which is Energy multiplied by Time. Speaking of this, General Smuts says*: 'The physical world thus becomes at bottom structured Action, activity structuralised in bodies, things and events. Thus arises the apparent material universe which surrounds us, and in our bodies forms part of us.'

He goes on to say that there is more in bodies, things and events than is contained in their structures or material forms. The inner action transcends the outer structure and there is thus a trend in things beyond themselves. From this springs the conception of fields of force and the force of evolution.

* *Holism and Evolution*, by J. C. Smuts.

What is this 'inner action'? Is it something in the psychological realm such as a mental influence? Can something on the mental plane be transformed into something physical?

In what way does matter come into existence? What is the origin of the forms of energy known to physics and chemistry, such as heat, light, magnetism and electricity? There are textbook explanations showing that energy is released by internal changes in the atom; but whence comes the immense force residing in the atom, and why if some structural change occurs inside the atom should there be a change of energy? Is there some intimate connection between structure and energy?

And how does structure arise? What forces, for instance, are responsible for the beautiful and varied patterns of crystallisation? Fifty years ago no one doubted that crystallisation could be explained by the action of known chemical laws, but now chemistry alone seems inadequate to account for it.

Chemistry and physics are littered with such unsolved problems, and when we come to biology the weakness of modern science to explain anything basic is extreme.

'The intellect', Bergson wrote, 'is characterised by a natural inability to comprehend life.' So much is happening behind the scenes where the phenomena of life are concerned that biological processes seem to us to happen by magic.

Writing of the development of a fertilised ovum, Sir Charles Sherrington says*:

> The little ball (the germ-cells) can be likened crudely enough to a set of magic bricks. The one cell, the original fertilised cell, grows into two, and so forth. When that has gone on in the aggregate some forty-five times, there are twenty-six million million magic bricks instead of one. That is about the number in the human child at birth.
>
> They have arranged themselves into a complex which is a human child. Each cell in all that more than million-fold population has taken up its right position. Each has assumed its required form and size in the right place. . . . Each cell has taken on the shape which will suit its particular business in

* *Man on His Nature*, by Sir Charles Sherrington.

the cell-community of which it is a member, whether its skill is to lie in mechanical pulling, chemical manufacture, gas-transport, radiation-absorption, or what not.

More still, it has done so as though it 'knew' the minute local conditions of the particular spot in which its lot is cast. . . . It is as if an immanent principle inspired each cell with knowledge for the carrying out of a design. And this picture which the microscope supplies to us, conveying this impression of prescience and intention, supplies us after all, because it is but a picture, with only the static form. That is, but the outward and visible sign of a dynamic activity which is a harmony in time as well as space. 'Never the time and the place and the "agent" all together.' Here all three and always, save for disease. And dominating forces as diverse as powerful.

Notice the words 'knew', 'skill', 'prescience' and 'intention'. Notice too the phrase 'harmony in time as well as space'.

Here is a feat of engineering which finds no parallel in the works of man. Similar intelligence and skill which seem to operate beyond the limitations of time and space may be observed in the regulatory processes of Nature—in the balance of the blood, the healing of wounds and in many other directions too numerous to mention.

Confronted by such miracles, we are like spectators in a cinema who have no knowledge of cinematography. We watch a drama unfolding on the screen, and perhaps we know just enough to realise that there is a projector which would correspond to our knowledge of chemistry. We say there are powerful catalysts such as enzymes, or chemical agents known as organisers and evocators which regulate the building processes in Nature. We marvel at the minute specks of protein called genes which, so the biologists tell us, carry in their chemical structure the hereditary experience of the race for untold millions of years, and much more besides. But surely this is straining the resources of chemistry too much? A suspicion is gaining ground even in the most august circles that there must be something *behind* chemistry. As Professor C. H. Waddington says*: 'The developmental side of biology—embryology, genetics and evolu-

* *Organisers and Genes*, by Professor C. H. Waddington.

tion—seems to be reaching a point where radically new types of thinking are called for.'

There is much more in cinematography than the projection onto the screen. There is the script writing, the casting, the shooting, and so forth, all of which is invisible to the audience. So too there are a great many unseen things behind the visible phenomena of chemistry, and in life processes any attempt to account for what happens by chemistry alone is futile.

Those who criticise the mechanical school of thought say, 'You have left something out of your calculations.' The mechanists reply, 'What we have left out is beyond the scope of science and therefore cannot be discussed.'

The same reluctance to penetrate deeply enough is still more apparent in the current views on evolution. Professor Bateson says*:

> The many converging lines of evidence point so clearly to the central fact of the origin of the forms of life by an evolutionary process that we are compelled to accept this deduction, but as to almost all the essential features, whether of cause or mode, by which specific diversity has become what we perceive it to be, we have to confess to an ignorance nearly total.

Reverting to our cinema film, we can see that the origin of it is a single Idea in a man's mind. It is then wholly mental and fluid—in fact potential. Later on it begins to pass from the mental to the material plane in a series of jumps or stages. The script is written, financial arrangements are made, the actors are cast, studios are booked. At a later stage the cast and the technical staff assemble and the shooting begins. Finally a great many films are reproduced from the original one, and these are shown repeatedly to millions of cinema-goers.

From the single idea to the many showings there is thus a progressive fanning-out process in time, space and number, a passage along another dimension which is not time but something which we can call the Fifth Dimension. The finished film exists potentially in the idea; then in successive stages an organisation consisting of a series of structures is formed into which the idea flows until it is materialised as a succession of showings

* *Problems of Genetics*, by Professor Bateson.

in chains of cinemas. This 'lives' until there are no further possibilities to be actualised, no more people who want to see the film. The film then 'dies'.

Perhaps this gives some idea of how the creative process in Nature takes place and what governs life and death. We can imagine everything beginning as an Idea and then progressing step by step through the Fifth Dimension into the visible world. Science studies the last stages of the process, from the point where the mental has already passed into the material. To a great extent it examines effects rather than causes; and because it is looking at the 'ends of an open fan' it fails to see the unifying principle and tends to regard all things as separate.

This is peculiarly characteristic of modern thought. There is reason to suppose that in past ages mankind saw things as much more connected. The great religions insisted on a higher unifying principle. People felt themselves closer to one another and to Nature. Art, science and religion were one. Sympathetic magic was believed in. But today with our ever-increasing specialisation and our worship of external factors, our thought, faith and way of living are becoming fragmented.

What we need are some new discoveries a little nearer to the 'hinge of the fan', knowledge which throws light on the region where the mental is passing into the material. Men of imagination have long realised this. Maurice Maeterlinck says*:

> It seems more certain that, as the cells of an immense organism, we are connected with everything that exists by an inextricable network of vibrations, waves, and influences, of nameless, numberless and uninterrupted fluids. Nearly always, in nearly all men, everything carried along by these invisible wires falls into the depths of the unconsciousness and passes unperceived, which does not mean that it remains inactive.

What are these vibrations, waves and influences? Are they, as the materialist thinkers maintain, something outside the province of physics, or can some of them be investigated in the laboratory?

These are some of the questions which this book will try to

* *The Unknown Guest*, by Maurice Maeterlinck.

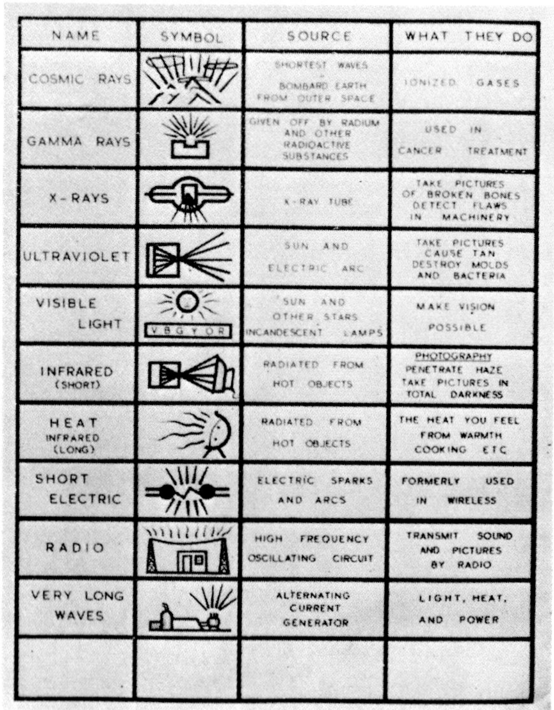

NAME	SYMBOL	SOURCE	WHAT THEY DO
COSMIC RAYS		SHORTEST WAVES BOMBARD EARTH FROM OUTER SPACE	IONIZED GASES
GAMMA RAYS		GIVEN OFF BY RADIUM AND OTHER RADIOACTIVE SUBSTANCES	USED IN CANCER TREATMENT
X-RAYS		X-RAY TUBE	TAKE PICTURES OF BROKEN BONES DETECT FLAWS IN MACHINERY
ULTRAVIOLET		SUN AND ELECTRIC ARC	TAKE PICTURES CAUSE TAN DESTROY MOLDS AND BACTERIA
VISIBLE LIGHT	V B G Y O R	SUN AND OTHER STARS INCANDESCENT LAMPS	MAKE VISION POSSIBLE
INFRARED (SHORT)		RADIATED FROM HOT OBJECTS	PHOTOGRAPHY PENETRATE HAZE TAKE PICTURES IN TOTAL DARKNESS
HEAT INFRARED (LONG)		RADIATED FROM HOT OBJECTS	THE HEAT YOU FEEL FROM WARMTH COOKING ETC
SHORT ELECTRIC		ELECTRIC SPARKS AND ARCS	FORMERLY USED IN WIRELESS
RADIO		HIGH FREQUENCY OSCILLATING CIRCUIT	TRANSMIT SOUND AND PICTURES BY RADIO
VERY LONG WAVES		ALTERNATING CURRENT GENERATOR	LIGHT, HEAT, AND POWER

Fig. 1. A brief classification of Electro-magnetic Radiations

answer. We shall begin by trying to get a clear idea of what science already knows about radiations of different kinds, and after this we shall cross the frontier and see what lies a little nearer to the Source.

CHAPTER TWO

RADIATIONS KNOWN AND UNKNOWN

> The life of every brute and of the plants is drawn from compounds having potency, by the ray and movement of the Sacred Lights.
>
> DANTE (trs. P. H. Wicksteed)

MOST of the radiations which figure in the scientific text-books are represented as forming part of the Electro-magnetic Spectrum, which is usually shown as a graduated table of all the known frequencies or rates of vibration.

Roughly speaking there were formerly two views as to how these energies travel. The first favours the conception of undulatory or wave movement through an all-pervasive medium called the ether. This accounts for all the vibrations except those of sound and ultra-sound which of course travel in air. The other view is that the source of energy emits fine particles often referred to as quanta, travelling at very high speeds through space.

The text-books of both schools of thought describe radiations with an enormous range of frequencies, or number of vibrations per second, from around zero up to astronomical figures.

The different radiations are tabulated diagrammatically according to their frequencies in Fig. 1.

We must now make another classification of these radiations known to science by listing the text-book references to charged particles. Some of these are shown in Fig. 1. The behaviour of each particle depends upon its charge and its weight or mass.

Present-day atomic theory does not provide a picture of the atom which is easy to visualise. However, an early theory of Rutherford gives us a working model in which the positively charged nucleus may be considered as a sun, and the negatively charged electrons which surround it as planets. In this picture the inner electrons have many times the energy of the outer electrons because of the immense speed with which they revolve. The up-to-date picture of the atom, provided by Schroedinger, is a wave picture where the inner electrons have an extremely high frequency and the outer electrons a lower one. These two pictures between them give us much information about the internal behaviour of the electrically charged particles which constitute the atom.

FIG. 2. SOME ELEMENTARY, HYPOTHETICAL AND COMPOUND PARTICLES

		Comparative Mass
Electron	Beta particle	1
Proton		1,878
Neutron		1,878
Meson	Pi, Mu, etc.	Varies considerably
Positron		1
Neutrino		Zero
Photon	Light quanta, X-rays, Gamma-rays	Zero
Alpha particle		4 × 1,878 (approx.)

It is possible to accelerate charged particles to such velocities that when they strike atoms, electrons may be ejected from their orbits. Free electrons in the vicinity of the atom will replace the missing electrons and the atom will be restored to its original state, energy being released in the process.

For some time it was thought unlikely that any radiation could exist with a higher energy than Gamma-rays, but at length physicists began to notice that something unaccountable was happening to their electroscopes. No matter how carefully they screened them with lead, the electroscopes persisted in losing some of their electrical charges.

This looked as if an unknown form of radiation was passing through the lead, which is a powerful absorber, and altering the electrical balance of the atoms. Could it be coming from the Earth? The electroscopes were sent up in balloons, but the higher they rose the greater was the leakage. At a height of 100,000 feet it had increased 200 times over. Evidently the radiation was coming from outer space.

With his associates, Professor Millikan put in a very great deal of work in investigating it, and today we know that these so-called cosmic or Millikan rays will pass through 1,500 feet of ordinary soil, or more than 3,000 feet of water. Since the Universe is mostly empty space, they may travel for millions of years before encountering any solid matter.

Year in, year out, they fall upon the Earth. Every second about six 'shots' pass through an area about the size of a man's head, and perhaps they play some important part in the processes of life. But what their origin is and precisely what they are, no one knows.

Are there any other forms of radiation, radiations which are not listed in the text-books? It will appear in the course of this book that all forms of matter emit radiations.

Where living matter is concerned, persons possessing a certain kind of sensitivity, or an extra sense of perception, can become aware of emanations emitted by human beings or animals. Perhaps when we come down the evolutionary scale to insects, molluscs and plants the emanations are not so obvious, though they are clear enough in the case of glow-worms and fire-flies which happen to emit radiations falling within the visible lightband. Many of the mysteries of Nature, such as the migration of birds, might be explained by the presence of radiations which are received in some unexplained fashion.

The scientists Gurwitsch and Frank have recorded what they call 'mitogenic radiation', which is said to come from the growth process of living cells, namely cell division in the roots of growing plants; and more recently in Colombia University I. I. Rabi, P. Kusch and S. Millman, using a new kind of apparatus, gave scientific proof of rays or vibrations which pass between one molecule and another. They showed that each cell, living or inert, is a tiny radio transmitter and receiver which gives a continuous broadcast, and that the waves range in length over the whole of the Electro-magnetic Spectrum. Indeed, they found that the jumble of these vibrations emitted is almost incredible. A single molecule, they say, may give off rays of a million different wave-lengths but only on one frequency at a time.

There is reason to believe that some of the ancient races were familiar with this idea of radiation from living creatures; but this knowledge seems to have been lost and re-discovered after the lapse of millenniums. In more recent times it was Goethe who first suggested studying the vibrations emitted by the human body. Newton regarded the Universe as a dead mechanism, but Goethe with his poetic insight saw it as filled with life and colour.

About a quarter of a century later, Goethe's idea was taken up by another famous German. Baron von Reichenbach was one of the most distinguished chemists of his day and the discoverer of creosote and many other chemical compounds. But his heart was not in coal-tar, it was in the curious experiments which had been made by Mesmer. Towards the middle of the nineteenth century he began to carry out experiments of his own in this direction, using as his subjects people who obviously possessed some form of extra-sensory perception. Many of them suffered from various kinds of nervous disorders, but some were in good health. He called them 'sensitive subjects', and with their aid he accumulated a wealth of evidence indicating that there existed in Nature a peculiar kind of energy which he named the Odic Force. He found it in magnets, in crystals, in light, heat and living cells; he found it manifesting itself wherever any kind of chemical reaction was taking place.

This energy could be accumulated or conducted along wires, focused by a lens, or distorted as a candle-flame is distorted by blowing on it. It seemed to be a vital energy since he discovered that certain people could transmit it to others—for healing, the relief of pain, or even to produce anaesthesia. In 1844 Reichenbach published his *Researches* and the book created a considerable stir in this country. But doctors and scientists were sceptical and could not accept his claims. Yet the great Humboldt said of his work: 'The facts are undeniable: it now becomes the task of Science to explain them.'

Even more scorn was heaped on the head of the American doctor, Abrams, who held the view that all matter emitted radiations and that the human body could be used as a receiver for them. Each of the organs, he found, acts as a separate broadcasting station, emitting waves of a certain frequency when not diseased or weakened. He drew up a table of frequencies which he believed represented the correct vibrational rates of the various organs. Disease, he said, was an altered rate of vibration. He claimed to be able to correct wrong vibrational rates by means of a device of his own invention, and he also claimed to be able to diagnose a patient's physical condition from a specimen of his blood.

Doctors regarded him with the greatest misgiving, and in the medical world the term 'Abrams' Box' is still a term of scorn and derision. Abrams' great mistake was to be somewhat bombastic about his powers to diagnose, so that he appeared in a ludicrous light when he was tricked in a test in which rabbit's blood was substituted for human blood.

The work of Abrams was carried on by Ruth Drown, of California, who in 1939 published an account of her work in *Theory and Technique of the Drown Radio-Therapy and Radio-Vision Instruments.* She developed instruments for diagnosis and treatment and found she was able to treat patients through a 'link' such as a blood specimen. Her work still continues in America, though it is not officially recognised in medical circles.

In England Abrams' work was taken up by Dr Guyon Richards, who lived in Lister's old house in Park Crescent, London. Also by Drs Dudley Wright, Ernest Martin and others. Dr Richards was much impressed by the results achieved by homeopathy, a therapeutic system which consists in administering so-called 'potentised' dilutions of minerals and herbs.

Hahnemann, the originator of homeopathy, was ridiculed by the medical profession of his day. How, it was asked, could one-millionth of a grain of belladonna possibly cure scarlet fever? He and his followers knew that the almost magical effects of these microscopic doses depended upon potentisation. Some subtle form of radiation entered into the dose. If you are an allopathic doctor, you pin your faith on chemistry and administer doses of drugs, but if you are a homeopath you diminish the chemical action and rely on the radiative force in the potentised dilution.

At the moment of going to press there is news that fifteen years of research work carried out by the Boyd Medical Research Institute in Glasgow has shown that this potentisation is a reality. In the experiments a crude drug was added to a diluting fluid in the proportion of one to ninety-nine drops. After vigorous shaking, a single drop of the diluent was added to a further ninety-nine drops, and so on.

When this process had been repeated thirty times the fraction of the original drug remaining was about one divided by an integer with sixty noughts. Not one molecule of the drug remained, yet more than 1,000 carefully controlled experiments proved beyond any doubt that these high dilutions will affect living cells.

Dr Richards thought that this radiative force in homeopathy was the same force which had been discovered by Reichenbach, Abrams and others. He improved on Abrams' work and tried to prove that his table of rates bore a definite relationship to the table of atomic weights. He also confirmed Reichenbach's discovery that the human body radiates a form of energy which had not hitherto been detectable by instruments. Towards the end of his life he formed the London Medical Society for the Study of Radiesthesia and began to make use of a pendulum for diagnosis. This is 'dowsing' applied to Medicine.

What is water divining, or dowsing? *The Divining Rod* by Sir William Barrett and Theodore Besterman, an authoritative book published twenty years ago, describes the work of some of the best-known dowsers and proves beyond doubt that water divining is a genuine art. Not only can water be discovered by its aid: minerals, metals, archaeological remains and many other things have been traced in this way.

Some people believe that the rod with which Moses struck the rock was a divining rod and the term 'Mosaical rod' occurs in literature. Ancient Egyptian bas-reliefs show diviners at work with bells held in their extended hands. In the reign of James I a certain John Scott 'who pretended to the use of the Mosaical rods' was engaged by David Ramsey, a clock-maker, to dowse for buried treasure in the cloister of Westminster Abbey. Unhappily he failed, but the chronicler records: 'The true miscarriage of the business was by reason of so many people being present at the operation; for there was about thirty, some laughing, others deriding us.'

This looks as if our ancestors regarded dowsing as a psychic operation, and in fact the authors of *The Divining Rod* suggest that knowledge of the object searched for enters the operator's subconscious mind and usually shows itself by a slight involuntary twitching of his muscles. Some diviners even believe that the gift is telepathic, but it seems more likely that the dowser's neuro-muscular system is affected by this same radiation which is emitted by all forms of matter and that his mind acts as a selector or tuning-in apparatus. Some dowsers get a reaction not only when they are immediately over the underground stream, or whatever it may be, but at regular intervals away from it—rather like a series of diminishing ripples.

This hardly seems like the work of mind, and yet mind enters into it in no uncertain way, because in radiesthesia, of which dowsing forms a part, we are passing beyond the frontiers of matter into another sphere where mind is a demonstrable force. So when a dowser sets out in search say of coal, he must have coal in his mind; or if he prefers, he may take a sample of coal which acts as a 'witness' or link. It not only helps him to focus his mind on 'coal underground', but being coal it emits radiations similar to those of the substance he is seeking. Some dowsers find samples more effective when they hold them against the solar plexus.

Since dowsing works with inanimate matter it is not surprising that it can be used to detect the still more powerful radiations from living substances. Just as dowsers have discovered concentric rings of reaction at regular intervals from say, an underground stream, so they have found circling the body of a patient similar rings which vary with his state of health.

Most medical dowsers use miniature pendulums consisting perhaps of a small piece of ivory or amber suspended on a length of thread, and these pendulums are found to gyrate when placed in one of these rings emitted by the patient, the manner of the gyration giving a clue to his state of health.

Pendulists also make use of samples of diseased tissue, bacterial cultures and so forth which emit their own particular radiations, and by various techniques they attempt to diagnose the cause of the trouble. They are also able to check any proposed remedy, whether homeopathic or otherwise, by placing a sample of it near the patient and observing if its radiations bring the gyrations of the pendulum nearer to normal.

There is no doubt that medical dowsing works and that some of its practitioners acquire considerable skill, but unfortunately unless the operator is exceptionally capable and experienced the personal element may interfere with the results.

It may be that there has been too much loose talk about 'human vibrations', 'personal radiations', and so forth on the part of people with little or no scientific training. However it may be, doctors are suspicious of such ideas and medicine is apt to lean rather heavily on chemistry. But chemistry as we have said earlier is not very far back in the line of causality. What lies behind it?

Here and there even among the most orthodox you find an admission that there may be something more at work than chemistry. Joseph Needham, who cannot be accused of undue credulity, says*: 'It is possible to picture a single molecule or molecular aggregate (perhaps of a para-crystalline nature) exerting an influence around itself in all directions of space for a considerable distance, even into microscopic dimensions.'

There are in fact formative forces in Nature which cannot be pin-pointed in the Electro-magnetic Spectrum. E. Pfeiffer has produced patterns of some of them in rather the same way that many of us produced patterns of magnetic forces in iron filings when we were at school.

He covered a smooth plate of glass with a thin film of copper chloride which crystallises when it cools. By adding the juices of various plants, human or animal saliva, urine, or diluted blood, he found that the radiations from these substances draw characteristic designs in the crystals. Lily juice, for instance, draws a sort of 'lily picture', while agave produces a strange design which appears prickly.

Blood, he concluded, possesses a strong radiative energy and each person has his own peculiar blood crystallisation pattern. By using sodium sulphate instead of copper chloride, he found that those forces which build up and maintain the shapes and proportions in Nature depend to a great extent upon radiations from the Sun. The patterns are much less distinct at night or during a solar eclipse. His latest work seems to indicate that crystalline formations obtained from certain solutions can be affected by thought processes.

We must also take note of the work of J. C. Maby who with T. B. Franklin and others tried to record radiesthetic vibrations with delicate electronic apparatus. He found that his instruments

* *Order and Life*, by J. Needham.

were upset by oncoming gales, thunderstorms, snowstorms, fog, heavy clouds, and also by near-by moving objects such as aircraft, cars and human beings; but in these latter cases only when they were at certain critical distances, suggesting a parallel to the 'rings' known to dowsers.

Among other men who have contributed something towards our subject are George Crile, an American doctor, and Georges Lakhovsky.

Crile, as others have done, formed a view that the human body is an extremely complex electronic apparatus in which both short-wave and long-wave radiation play important roles. Each living cell he saw as a tiny electric battery generating its own current by chemical action, and the nervous system as a network of highly specialised electrical conductors.

Lakhovsky developed the study of radiations from living matter on rather different lines. He regarded each of our 200 quintillion living cells as a tiny battery with the nucleus acting as an oscillating circuit and giving off radiations. In disease the frequency changes.

He drew an interesting analogy between human bodies and radio sets. Our radio sets work because our aerials are placed in a variable electro-magnetic field created by the waves emitted by the broadcasting stations. The cells of our bodies work in a similar way. They move rapidly with the Earth's rotation and revolution through a variable electro-magnetic field generated by radiations from the Sun, the Milky Way and the distant galaxies. Like all living matter, they are balanced on a knife-edge, for they are constantly under the influence of radiations of higher and lower frequency which may compel them to alter their vibrational rate and so produce one of the many diseases whose root cause is so baffling.

Lakhovsky quoted Albert Nodon's experiments which showed that the radiation from plants is as great as that of radium or uranium. That of insects is from three to fifteen times greater, while the vital cells in human bodies radiate still more powerfully. He showed that every living thing is a delicate and complex radiative apparatus depending upon chains of interlocking processes within and supported like a fish in water by a sea of vibrations without. In his view the whole Universe is knit together by a 'universal plexus of cosmic rays', and all forms of matter, ranging from rarefied gases to solid rock are simply varying degrees of condensation in this plexus. In radiation everything lives, moves and has its being.

Let us now follow the work of another pioneer who has spent many years of his life investigating radiations, particularly as they appear in life processes.

This will involve a journey into very strange country, a country which lies beyond the frontiers of text-book science. It may not be easy to describe what has been discovered there because we shall be looking at it through the aperture, so to speak, of a series of laboratory experiments which may be compared to aerial photographs taken by an explorer flying over uncharted territory. The task of interpreting these photos, of taking a great many more, and interpreting these too may take 100 years. The most we can hope for is a fragmentary pre-view.

In earlier days when an explorer reported having seen nuts the size of a man's head with hair on them he was laughed to scorn. Europeans could not imagine a coco-nut. Perhaps we shall have difficulty even in imagining some of the things described in this book, but that is no reason why they should be rejected. Any Victorian scientist who had been bold enough to forecast radar or television would have been derided as an irresponsible visionary. In these days it is as well to preserve an open mind.

CHAPTER THREE

SOUND AND ELECTRICITY

There's not the smallest orb which thou beholdest
But in his motion like an angel sings
While quiring to the young eye'd cherubim
Such harmony is in immortal souls;
But whilst this muddy vesture of decay doth grossly
 close it in
We cannot hear it. SHAKESPEARE

It is necessary first to say a few words about the man who initiated these experiments. Science is impersonal, or should be; it is the discoveries which matter, not the men who make them. Nevertheless, when it is a question of revelations which are not in line with established tenets of science, the personal factor cannot altogether be excluded. Readers will wish to be assured that they are not hearing about the work of an impracticable visionary who has never passed through a hard school of training or held a responsible post.

George de la Warr, who was born in 1904, passed the Associate Membership examination of the Institution of Mechanical Engineers at the age of twenty, and three years later that of the Institution of Civil Engineers. Later he qualified as an Associate Member of the Town Planning Institute, since when he has served as Chief Engineering Assistant of an oil refinery, Chief Constructional Engineer of the Firestone Tyre and Rubber Factory, and also Development Engineer in the same firm. For five years he was building roads and bridges for the Somerset County Council, and for sixteen years the Oxfordshire County Council employed him as their Chief Engineering Assistant.

In 1929 he married the daughter of a scientist, a lady who proved exceptionally capable of helping him with the work which he was to take up later on.

As a young man he had been very much interested in medical science, and particularly in the conflicting views held by allopathic doctors and the homeopathists. He knew from practical experience that homeopathic methods worked, but he felt there was something important behind homeopathy which needed investigation. Modern medicine had been in existence for only a very short time, but the human race had not died out. Gradually he became convinced that Nature held the remedy for most human and animal ailments and that it could be applied particularly by the judicious use of plants and minerals. This, however, was not quite the line that he eventually followed.

It seemed to him that medicine should make use of the curative radiations in Nature rather than relying quite so much on allopathic drugs. Perhaps these radiations could be produced synthetically. But how?

One day a friend remarked casually that each plant probably emitted its own particular musical note. This gave a new turn to his thoughts.

This 'music' might be ultrasonic, in which case it would be inaudible to human ears, or it might be a harmony in some other kind of radiation. Was it possible to pick it up with a small receiving aerial? He experimented with a crystal set and was disappointed to find that it recorded nothing at all. On the other hand he had some slight success when he tried a pendulum. But he says: 'The gyrations of this annoying instrument misled me so often that I wonder I persisted.' He says too: 'I also tried another form of radiation detector, a water diviner's twig, but I found this equally annoying—especially as my wife proved to be a better diviner than myself!'

However, this idea that living things such as plants emit musical notes led to some peculiar discoveries which linked acoustics with the electrical forces in Nature.

In the last chapter we saw that every living entity is a complex electrical apparatus composed of myriads of cells, each one a minute galvanic battery, the aggregate of which is capable of producing a considerable electrostatic potential.

Using a sensitive spot galvanometer capable of measuring minute electric potentials, de la Warr performed some experiments with plants. Dr J. Bose did a great deal of work with similar apparatus and was able to prove the sensitivity of plants to extraneous influences such as the passing of the sun behind a cloud. De la Warr confirmed some of Bose's findings, but he also found that a plant's electric potential was affected when the plant was treated with sound waves, or when he or his wife held their hands near it.

He also suspected that when the plant was rotated into a certain position in relation to the north-south axis of the Earth there was a very gradual change in the plant's potential. Could there be some fundamental relationship between the Earth's magnetic field on the one hand and the static electricity of the plant on the other?

Eventually he came to the conclusion that the cells in the plants were emitting some form of radiation which was stimulated by sound, and that this radiation was related to the static electricity generated in these cells. If then, he thought, plants were treated with sound waves of just the right kind, their cells should be stimulated sufficiently to bring about an increase in growth.

He tried an experiment with six potted plants, using an apparatus consisting of a number of variable resonators to produce the necessary sound waves. He realised that each type of plant would have its own special electrical conditions and therefore it would need a corresponding mixture of sound waves to stimulate its growth. We shall see how he arrived at the right mixture in a moment.

When the plants were correctly treated in this way he found they grew much faster than similar plants which had not been treated. In fact, the results were so striking that he decided to see if seed could be germinated out of season. So in October he planted six nasturtium seeds in a shallow bowl of earth and gave them a 'concert' twice daily. Within four or five days they had germinated and shoots were showing above the soil.

How was he able to choose the correct 'musical programme'?

Before answering this question we shall have to understand two things which figure rather prominently in this book. The first is 'wave-forms', and the second is something which is familiar to musicians, namely 'resonance'.

If you strike a note on a piano you get a simple vibration in air. Now if you strike another note which has the necessary harmonic relationship with the first note, the two will combine to form a chord, and so on. Sound waves are governed by a Law of Harmonic Relationship which is the very basis of music.

As we have seen, there are many other kinds of vibrations, such as light waves, radio waves and cosmic rays, which are vibrations in the so-called 'ether'. Each of them has a definite wave-length and is analogous to a single note on the piano. And like sound waves that combine under the Law of Harmonic Relationships. They will also combine if they have the right harmonic relationship.

To illustrate this let us take three different wave-lengths, A, B and C. If we combine them we shall get a complex wave pattern. Now let us add a fourth vibration and we shall again get a change of pattern. If we keep on adding vibrations which have the necessary rhythm and harmonic relationship, we shall build up a steady, harmonious pattern. In fact, we shall create what is called a wave-form. This is seen, or rather heard, clearly in music where single notes combine to form chords, and chords combine to form harmonies.

Once again, if we strike a note on a piano it will vibrate in harmony with parts (or components) of a note higher or lower in the octave. This is called resonance.

In the same way one radio wave will resonate with another radio wave. A radio set tuned to 265 metres is in resonance with the Light Programme transmitter which is emitting waves of the same frequency. If we wish, we can regard resonance as a sort of affinity such as is found in various chemical elements, but an affinity which takes little account of distance.

It was at this stage in his experiments in the year 1942 that de la Warr had that flash of inspiration which was to enable a logical sequence of experiments to be made at intervals over the next few years. Knowing that in radar it was necessary to relate the length of the aerial to the signal wave-length, he made up some small stub-aerials of different lengths. The method he used was to take a small piece of brass sheet as a base-plate and mount several short masts on it. To each mast he fitted a short piece of copper wire to act as a stub-aerial. Taking a pair of pliers he kept shortening these aerials until the necessary relationship was arrived at between their lengths to give resonance with a near-by plant.

He thought that if the plant was emitting a complex wave-form he might be able to detect it with a complex arrangement of aerials. The resulting signal would then be detected in the base-plate.

It proved to be a reasonably satisfactory piece of detection apparatus, but it was a laborious process getting the aerials the right length. Some easily operated method of determining the moment of resonance was needed, and it was this problem which caused de la Warr to make his portable detector.

A photograph of it and also a diagram of it in cross section are shown in Figs. 3 and 4.

You will notice that it has a resonant cavity and a metal plate which also acts as a resonator. (Imagine a cavern in which if you pitch your voice correctly you will get a reverberant effect.) An 'antenna wire' connects the plant or seed to the stub in order to introduce the plant's radiations into the cavity. This simple piece of apparatus was found perfectly effective for detecting resonance when the plant responded to the correct musical programme.

But how is it known when there is resonance? This is where the third kind of radiation comes in, a radiation emitted by the operator.

On the top of the instrument is a thin rubber diaphragm. The operator strokes this diaphragm, lightly passing his finger repeatedly from right to left, and when there is resonance this is shown by a tell-tale crackle and a sticky feeling in the rubber known technically as a 'stick'.

If we follow up what is happening and why this stick proves that there is resonance, we shall have to cross the frontier which divides the world of matter from another world where mind is an operative force. For as we shall see later, the mind acts as a modulator of complex and continually changing wave-forms which are emitted from the human body. To hold in mind a steady thought of something is to attune your bodily radiations to those of that something.

In order, then, to tune-in to the plant's radiations the operator focuses his mind on the plant. The wave-forms which he is emitting are then in a harmonic relationship with the plant's radiations and this reinforces the effect. If there is resonance

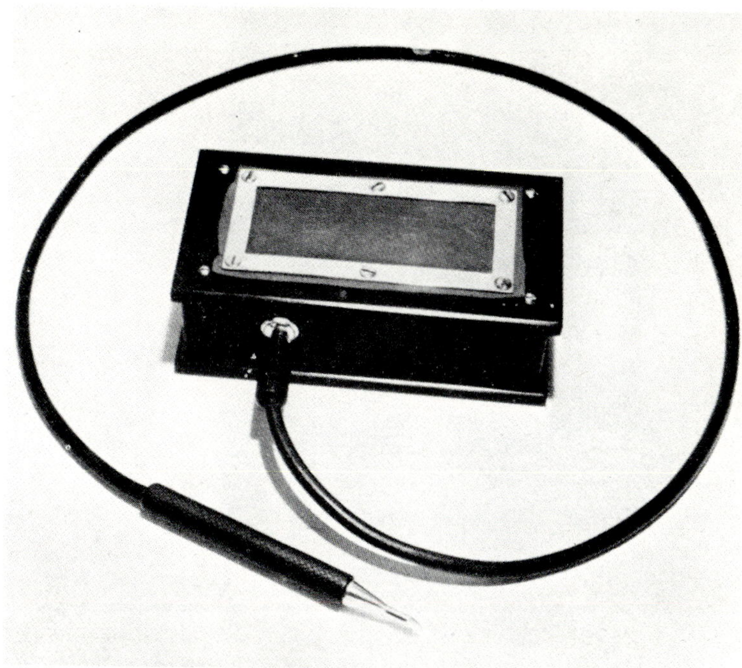

FIG. 3. Portable Detector and Antenna

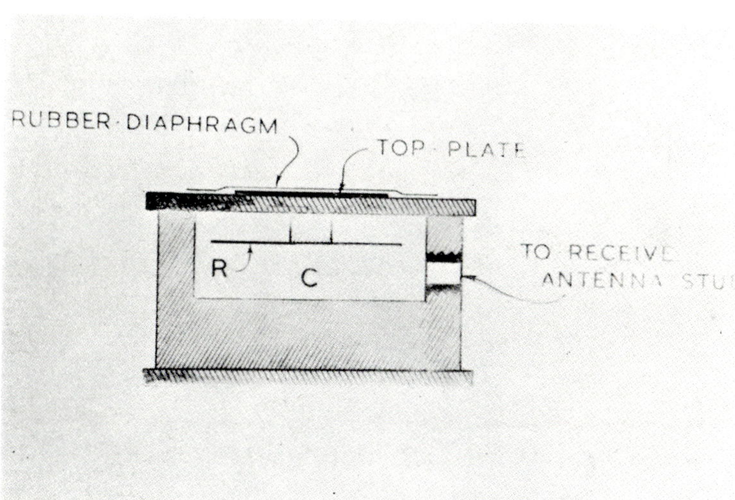

FIG. 4. Cross-section through Portable Detector

Fig. 5.
Diagnostic Instrument
(Sliding cursor in use)

Fig. 6.
Diagnostic Instrument
(Preliminary tuning operation)

between the plant and the subtle radiations associated with the musical programme, it is revealed by a crackle in the rubber diaphragm when the finger is passed over it.

We shall return to this later. It is not an easy thing to accept without adequate proof, and de la Warr only came to believe it gradually as weight of evidence in support of it accumulated. It has been confirmed over the years by a great many operators who were, and still are, using the instruments he invented; and finally by a most astonishing piece of photographic apparatus —a camera.

CHAPTER FOUR

NEW METHODS OF DIAGNOSIS

> The great bulk of people infinitely prefer the continuance of a problem which they cannot explain to an explanation which they cannot understand.
>
> LORD BALFOUR

HAVING established a relationship between these different kinds of energy and devised an instrument which in the hands of a capable operator would show when there was resonance between one radiation and another, it was only a short step to applying this new technique to the detection of disease radiations.

Lakhovsky regarded disease as an altered frequency. There were other pioneers who affirmed this and who had invented appliances to pick up the disease vibrations and match them against the vibrations of healthy organs or diseased tissue for comparison.

The first record in the Patents Office of such an apparatus is an application filed in 1924 by Arthur Whiting, of Victoria, Canada, for 'Improvements in or relating to apparatus for use in the Diagnosis and Treatment of Disease'. Dr Abrams, Dr Boyd of Glasgow, Dr Guyon Richards, Ruth Drown, J. C. Maby, Wiglesworth, Starr White and others had developed diagnostic instruments and carried out experiments, but as yet radiesthesia could hardly be called a science. These pioneers had touched the fringe of something which was likely to lead physics in a new direction and perhaps eventually be of great importance to medicine, but further discoveries were needed to dispel the obscurity in which it was shrouded and to place it on a firm basis of verifiable fact.

Returning to the multiple aerials which de la Warr used to detect the complex radiations from plants, he found that instead of cutting each stub-aerial to a suitable length it was almost as satisfactory to have a number of pieces of wire all the same length arranged rather like a square harp. By tapping these wires at different positions it was still possible to get the necessary resonance with the plant.

In other words, this new piece of apparatus behaved rather like a violin when the fingers of the violinist are pressed firmly on the strings. By dividing each wire into ten equal parts it was possible to record repeatable positions in tenths. Thus when this harp-like device was tuned to detect the disease radiation of, say, a culture of the tuberculosis bacterium it always showed the following positions:

1st Wire	4/10			
2nd Wire		3/10		
3rd Wire			1/10	
4th Wire				0

These sliding contacts on the wires were not very practical, and de la Warr found that if a wire was curved into a circle the sliding contact could be operated by a knob. In fact, if a panel was made carrying dials which operated sliding contacts on wires fixed beneath it, an arrangement of dial settings could be found for each disease.

During the war de la Warr built a number of experimental instruments for the diagnosis of disease, incorporating nine dials, or resonators. These resonators were then set according to the 'rate', or arrangement of numbers necessary to produce the 'harmonious condition'.

The next thing to do was to fix the composite vibrational rates of each healthy organ, disease, and so on. Earlier inventors of diagnostic instruments had worked out tables of dial settings representing these rates. At first de la Warr tried merely to improve upon these existing instruments and he retained as far as possible the same arrangement of dials because he did not wish to start a second school of thought; but he soon saw that he would have to follow a line of his own and work out a new table of rates based on his own research work.

Some of the first disease rates to be fixed were determined from cultures of bacteria prepared by a friendly pathologist. The method was a simple one of trial and error. Taking, for example, a specimen of *Streptococcus Viridans* and placing it in the well of the instrument (Fig. 6), the operator, Mrs de la Warr, would focus her mind on *Streptococcus Viridans* and then turn Dial No. 1 until she got a reaction on the rubber diaphragm. The necessary setting proved to be 60. But this figure is common to the whole species of *Streptococcus* and further dial settings may be added to qualify the first one.

Leaving Dial No. 1 set at 60, she then turned Dial No. 2 until a second reaction appeared, corresponding to a further component of the wave-form of the disease. The second setting proved to be 5. The process was repeated with Dial No. 3 which she found had to be set at 2. Dial No. 4 gave no reaction, showing that the different rates of vibration in the wave-form had all been accounted for.

Thus the radiation from *Streptococcus Viridans* required three dial settings to produce the necessary synthetic wave-form which would resonate with it. In the table of rates it was entered as:

Streptococcus Viridans 60.5.2.

In the course of time more than 4,000 recognition rates were determined in this way and in practice they have nearly always proved correct. But the experimental sets were found somewhat unreliable in the diagnosis they gave. Some element of uncertainty was apparent: they needed a stabilising factor. And so de la Warr began to carry out further experiments to discover how this uncertainty arose.

After various attempts, it occurred to him that like Reichenbach he might try the effect of a bar-magnet. This turned out to be the correct key. By 1943 he had invented a simple magnetic device consisting of a bar-magnet standing upright with a knob—as shown between the two wells in Fig. 6—which rotated it like the steering column of a car. For some reason it was necessary to rotate it into a certain position before it would exert its stabilising effect, but when it was in this position it worked very well indeed. This magnetic tuner or stabiliser in spite of its extreme simplicity proved to be a most vital adjunct, a patent was applied for and it was incorporated in all the apparatus which he invented subsequently.

In the same year appeared his Baby Portable Diagnostic

which is now in use in many countries. Details of it are shown in Figs. 5 and 6.

There is a series of nine resonators each of which is connected to a common point, and at this point is a 'mixing platform' consisting of a thin sheet of metal where the energies from the resonators combine. In one corner is the rubber diaphragm. On either side of the magnetic stabilising device are two wells to hold blood specimens of the patient. We shall hear more about blood specimens later. At the present juncture it is enough to say that, as other pioneers had discovered, blood forms a radiative link between the operator and the person to whom it belongs. In fact, it serves as an admirable proxy for an absent patient.

On the left-hand side of the instrument is a hand-operated sliding cursor which passes over the detail sheets listing the various diseases, organs and other bodily constituents. Its function is to aid the operator in focusing his mind on whatever is necessary in the course of his diagnosis so that his controlled thought is made to play its part in the diagnostic process.

In 1945 when lecturing to the British Society of Dowsers, de la Warr demonstrated another kind of diagnostic device. This had a parabolic aerial to pick up the vibrations of a patient's cell-groups, which were then matched against synthetic vibrations set up in the instrument.

A man with catarrh stood in front of the aerial. The dials were set for catarrh, and as soon as Mrs de la Warr passed her fingers over the diaphragm a crackle was heard showing that there was resonance.

De la Warr himself had a touch of muscular rheumatism, and when the dials were set for this complaint and he stood in front of the aerial, there was again a crackle. Nor was there any doubt that these emanations from muscular rheumatism were some form of physical vibration, for he showed that they would behave like light waves. A mirror was set up in the lecture hall and when he walked down the centre aisle towards it they were reflected back to, and picked up by the diagnostic set. With any good reflecting surface the result is always the same, but a soft substance will not reflect these waves.

These instruments for diagnosis were very accurate in skilled hands but they did not quite rule out the personal factor. They were effective only if the operator was able to control his thoughts and form a clear mental concept of the diseased organ, or whatever it might be, with which he was dealing. In fact these instruments were no more than aids to personal skill, and although the help they afforded was great, de la Warr was not satisfied.

His ambition was to invent an automatic appliance which would eliminate the personal factor. This would not only overcome the difficulties of inept or poorly trained operators, it would lend itself far better to a scientific investigation of its merits. But as he was to discover, the invention of such an instrument was a very difficult matter indeed.

However, this research work into new methods of diagnosis was leading to some most interesting discoveries. There was no longer any doubt that an operator emitted some form of radiation which was a demonstrable reality. It seemed to emanate chiefly from the solar plexus since it was found that a sheet of perspex placed in front of the solar plexus partially cut it off. As far as could be judged, it was a very complex wave-form with a high degree of instability.

No doubt it was a mixture of many different kinds of radiation. De la Warr thought one component might originate in some property of the ether. Another was unquestionably a radiation the pattern of which seemed to change with every change of thought in the operator's mind.

This interaction between thought and radiation is not so incredible as it may appear. The science of ultrasonics on which such a vast amount of work has been done in recent years has discovered connections between ultra-sound waves and electrical pulses, and according to Sherrington and other biologists electrical pulses are intimately concerned with cerebration.

What are known as piezo-electric crystals will vibrate with extreme rapidity in an electric field and so emit ultra-sounds like the vibrating strings of a violin. And metals called magneto-strictors will stretch and shrink with similar rapidity in a fluctuating magnetic field, producing a similar ultrasonic effect.

The experiments with plants had shown that connections can exist between acoustic and electrical waves. Thoughts in the mind of the operator cause his brain-cells to send electric pulses coursing through his nervous system and these pulses alter the pattern of radiations which he is emitting.

There is thus a close analogy between what goes on in the human body and the behaviour of certain crystals and metals. This parallel between phenomena on different levels in the evolutionary scale is a most important one. The old adage, 'As above, so below', is one of those eternal truths which can be applied in almost every form of scientific research work which penetrates deeply enough beneath the facade of external appearances.

CHAPTER FIVE

INTO UNCHARTED TERRITORY

> The field is as much an integral part of matter as the sensible part which it surrounds. . . . The field may either be viewed as activities, or as structure—as elements of force, or as curves. Indeed, from many points of view structure and function, curve and force are convertible terms for the purpose of describing physical effects.
>
> *Holism and Evolution*, by J. C. SMUTS

SINCE it had already been shown that treatment with air waves stimulated the growth of plants, it seemed only reasonable to suppose that a physical effect would be produced if creatures higher in the evolutionary scale were treated in the same way.

De la Warr carried out a series of experiments on spiders and other creatures which proved beyond doubt that ultrasonic waves did in fact affect the behaviour of living cells. Cell fission could either be stimulated or inhibited according to the 'musical programme' which was projected at them.

Lecturing to the British Society of Dowsers in 1948, he said: 'It would appear that the human cell is sensitive to movement, and just as a crystalline structure will alter its polarity on distortion, so will a cell-group respond to air-wave bombardment and produce specific changes of electric potential resulting in nerve stimulus.'

Taking as an example the radiation pattern emanating from the mucous lining of the human larynx, he switched on three loud-speakers each emitting the appropriate note, and the resulting sustained chord of sound produced a dryness of the throat which was clearly felt by members of his audience. The frequencies used were 133, 264 and 455 cycles per second.

This phenomenon can be verified by anyone who wishes for proof. Any sensitive part of the human organism will respond to the correct combination of frequencies. For instance, the heart muscle is affected by a combination of 105, 160 and 214 cycles per second, while 98, 132 and 234 affect the speech area of the brain.

Sonic vibrations and the subtle radiations which they evoke can be potent for good or ill. Lord Horder said not long ago that we were dying in greater numbers from functional diseases such as coronary thrombosis, arterio-sclerosis and diabetes, and that in all these maladies noise played a part with its evil effects on our nervous systems. In America it has been discovered that the din of jet engines affects the lungs. Scientists who have studied the subject know that it is even possible to kill a man by an acoustic wave. In the first world war the possibilities of constructing a lethal sonic weapon were discussed.

It seemed to de la Warr that acoustic treatment sets were well within the bounds of possibility. There was no mechanical difficulty in producing air waves of different frequencies and combining them, but there remained the question of how to calculate the correct combination of waves to meet each individual case.

Suppose you wish to treat a case of tuberculosis. The recognition rate for Tuberculosis on the Delawarr Dials is 40.3.1. If the patient were treated with this rate it would stimulate the growth of T.B. cells. What is needed to inhibit their growth is the antidotal wave-form, or the 'key to fit the lock'.

It was found that this could be worked out by simple arithmetic, for it is the remainder of each dial marking beyond the original figure which gives the new figure that is being sought.

Dial No. 1 has markings from 0 to 100. The first figure in the recognition rate of Tuberculosis is 40. Subtract 40 from 100 and we get 60.

Dials Nos. 2 and 3 are marked from 0 to 10. Subtract 3 (the second figure in the recognition rate) from 10 and we get 7. Subtract 1 (the third figure in the recognition rate) from 10 and we get 9.

The antidotal rate for T.B. is therefore 60.7.9. If this rate is set up on an acoustic treatment set it will tend to inhibit the growth of tuberculous cells and so have a remedial action.

De la Warr designed several acoustic treatment sets for professional use and also some for self-treatment. From what has already been said, it is obvious that sonic waves of high intensity are dangerous unless used under skilled direction. So these latter instruments emit only waves of low intensity which cannot harm a patient.

Some of these sets were used by medical practitioners with excellent results. It was found that once the antidotal wave-form had been correctly fixed, the growth of bacteria was checked by treatment. Under the care of qualified medical men, even sarcoma and other grave maladies responded well.

These instruments were so successful that one of them was exhibited at the British Industries Fair in 1948. Tens of thousands of people must have visited the Fair during the fortnight when it was open, but few of them stopped to look at this strange little apparatus. No doubt the claims made for the curative properties of sound struck them as all too improbable.

After diagnosing a patient's physical condition with the instrument, acoustic treatment need not be prescribed if herbal, homeopathic or allopathic preparations are preferred. But here too a precise method of prescription can be used. The proposed remedy is placed in the well opposite to that which holds the blood specimen, and if it is an appropriate remedy it gives a reaction on the rubber detector. If it is wholly appropriate it has a localised effect, for as de la Warr found, there is resonance between the radiations of such a remedy and its *specific sphere of action* in the body.

This shows the connection between chemistry and radiations. A medical drug acts upon a patient because the radiations which it emits have the right harmonic relationship with those of the disease which is to be treated and the part of the body which is affected. A part of the wave-form emitted by the drug forms a discord with the radiations of the disease while another part of it is in resonance with the organ which is being treated.

Resonance explains many things which puzzle doctors and biologists. How is it that certain constituents of the blood such as hormones and enzymes, though minute in size, exert such a powerful effect, and moreover an effect which is highly localised? The answer is that they are exceedingly strong centres of radiation and that they act only on cells of the body whose radiations are in resonance with theirs.

Whence arises this homely lore of putting a dock leaf in one's shoe, or spurge on a wart? Why do some gardeners always plant two kinds of vegetable together and forestry experts encourage the growth of saplings by planting them in groups? Resonance is the key to much horticultural practice as well as to various enigmas of natural history.

De la Warr's instruments also showed connections between psychology and physiology. The astonishing discovery was made that it was possible to treat a patient's psychological condition as well as to diagnose it. It was almost unbelievable that somebody's mood could be changed in a very short time by turning a knob, but there it was. A test made on a whining infant showed that its whine could be cut off like music from a loud-speaker.

There was another case in which the diagnostic set registered 'anxiety', and at the same time it showed that both the heart and the supra-renal medulla were sluggish. When the patient was treated on the supra-renal medulla rate of vibration, his anxiety at once disappeared. The state of anxiety had depleted the action of these two organs (heart and supra-renal medulla) and had thus set up an energy-pattern which expressed 'anxiety' and which in time might have produced some 'disease of anxiety'.

This connection between pattern and psychology on the one hand, and between pattern and physiology on the other is most important.

There are some people who believe that emotions are equivalent to glandular secretions, but this is surely too materialist a view to meet the case. Emotion is in a different category, although glandular secretion profoundly affects it, just as a bang on the head or a shot of pentothal affects thinking. But what precisely is the bridge between psychological activity, such as thinking, sensation, emotion, and the physical body?

It seemed to de la Warr that everything whether animate or inanimate had certain properties in common with the magnetic field which is found in the immediate vicinity of a bar-magnet. If we imagine the numerous groups of cells in a human body acting as tiny transmitting stations, it is clear that they will emit an enormously complex network of radiative forces interpenetrating the body and intersecting at myriads of different points at short distances outside it. Such a network forms a pattern, or structure, and in general it is known to science as a force field. A suitable name for it in the case of a living creature is 'force field body'.

Was it possible, de la Warr wondered, to connect the force field body directly with sensation? Somewhere he had read that if a person were subjected to pain it would affect a compass needle. He decided to put this to the test.

Lying on a settee with his head in the direction of magnetic north and a small compass resting on his solar plexus, he persuaded his laboratory assistant to stick a pin into him. His wife who was anxiously watching saw the needle move; but was this due to his muscles twitching when he felt the prick of the pin? The experiment was tried again, this time with the compass suspended from the ceiling, and again the needle moved.

It was clear that sensation affected the force field body and so produced an electrical effect strong enough to be recorded by a magnetic instrument. So there was reason to suppose that the force field body was a bridge leading from laboratory phenomena such as magnetism towards sensation, which is in the psychological sphere.

It was also evident that these delicate instruments which had been invented were working on the level of the force field body rather than on that of the physical organism. As he gradually came to discover, the force field body is the breeding ground of good and bad health, and it is governed to a great extent by the psychological make-up of the subject as well as by solar and cosmic energy. We shall hear more later on about this cosmic energy and how it is evoked.

CHAPTER SIX

WORKING WITH A FREE HAND

> Obey the nature of things, and you are in concord with the Way,
> Calm and easy and free from annoyance;
> But when your thoughts are tied, you turn away from the truth,
> They grow heavier and duller and are not at all sound.
>
> D. SUZUKI in *A Manual of Zen Buddhism*

WHEN de la Warr came to Oxford in 1936 he was still very much interested in engineering, but as time went by he became more and more engrossed in the work which forms the subject matter of this book. Although he had to work long hours for the County Council, his researches and experiments made greater demands on him every month. They had to be carried out in his spare time. Often he would rise between four and six in the morning to draft out his ideas and plans for the next experiment.

His wife, too, as her two young daughters grew older had more time to give to this absorbing pursuit, and at length it became a whole-time job for her.

At first, not counting an instrument maker and a mechanic, they were a team of two, but as the years went by the task became too much for them. Their experiments with plants had shown that there was a connection between sound waves and electric potentials so that some electrical research work became necessary. They decided to engage a full-time physicist.

The amount of apparatus was increasing every week, and since Mrs de la Warr was carrying out experimental diagnosis and treatment, room had to be found for at least one patient. To house the physicist properly would mean obtaining permits from the Ministry of Works for an extension, and meanwhile they seemed to be on the verge of startling new discoveries. Their enthusiasm was too great for them to stomach irritating delays, and so after finding a physicist they persuaded him to occupy the gardener's cubby-hole! They chose him for his wide experience and his outstanding ability to design and construct experimental electronic circuits.

At first they carried out some of their experiments in the bath-room with the bath covered in, but it was most exhausting to have to remove dozens of bottles and pieces of apparatus each time somebody wanted a bath. In the course of time a small laboratory was built. When a little later a second laboratory was added, it seemed a hazardous venture buying more land and wondering how they were going to pay for everything, but they went boldly ahead, bought another two acres and included the cost in the mortgage on their house.

De la Warr's father, himself an engineer and architect of repute, who from the start had backed his son, gave him all the help he could afford. In order to be in complete control of their work they decided to keep expenses as low as possible and not seek financial help from outside the family. They had to be free to follow whatever lines of research work they wished to pursue and to develop new kinds of apparatus without interference from backers who might have different ideas from their own. Their researches were too much off the beaten track to appeal to the commercial instincts of the average financier.

Official backing of large research groups with narrow utilitarian ends in view is seldom suited to pursuing new lines of discovery. As Dr Alexis Carrel says*:

> It seems that the increased number of scientific workers, their being split up into groups whose studies are limited to a small subject, and over-specialisation, have brought about a shrinking of intelligence. There is no doubt that the quality of any human group decreases when the number of the individuals composing this group increases beyond certain limits. . . . They vainly exhaust their strength and spend their time in the pursuit of the conditions demanded by the elaboration of thought.

With a coherent flow of inspiration which came to them in great wealth and detail, all the de la Warrs had to do was to put it into practice. They suffered none of the frustrations and petty

* *Man the Unknown*, by Alexis Carrel.

irritations which come from Directions issued by superiors. Looking back, it is clear to them that these inspirations fell into an orderly pattern and produced new knowledge and new designs for apparatus which were always stepping-stones to their goal. Sometimes an inspiration would misfire, but as they would realise later, this was always due to their inadequate knowledge at the time which led them to draw wrong inferences from it. Invariably they were able to pick up the threads later on.

One of their first experiments in the new laboratory was to test a special kind of valve which they had been inspired to make some little time before. Originally it had been designed to work in circuit with a crystal receiver to detect radiations, but they had never completed it. This they now did, and when they fitted up a coil and a loop they found it was sensitive to light waves.

This was surprising. Broadly speaking, science knows no method of adapting two pieces of copper wire in such a way that when connected to another device made only of coils of wire and a simple Rochelle salt crystal it becomes sensitive to light. Their light-valve contained none of the normal light-sensitive substances, and yet for some reason light affected it. Evidently there was something important to be unravelled here, and de la Warr began to ask himself questions such as: Suppose we have a light-proof box in the side of which is a single pin-hole, what energy enters the box if light is admitted through the hole, say for one second? Is it possible that energy particles other than photons are trapped in the box?

They carried out a number of experiments with their valve, trying the effect of coloured light and the interaction of two or more coloured beams. And the valve registered strange phenomena which they could not understand. At the same time, the results as seen on the cathode-ray oscilloscope gave them their first fully scientific glimpse of the interaction of the radiations which fill all space.

Even after their new laboratory had been built, the space for working was all too small. Some of these experiments were conducted in a room no more than seven feet square, and they soon realised that when they used their new valve it built up peculiar conditions in this confined space.

There was an *atmosphere* which after about an hour affected their instruments. Their magnetic compass which at the beginning of the day's work had behaved normally would begin registering magnetic nodes in mid-air, and instead of pointing to magnetic north the needle would swing round when the compass was moved across the room. After about two hours this atmosphere became so strong that it would even dim the electric lights. They could stare directly at the filament of their 250-watt projector-lamp without blinking. Fuses would blow when the current was switched off.

At the time they could not fathom the reason for these peculiar happenings. The explanation came a few days later when they found that charged particles were being liberated, producing something like a particle cloud which affected a photographic plate and also caused a short-circuit across the terminals, blowing the fuse.

These experiments caused them temporarily to abandon sound waves and to concentrate on light waves, which led to entirely new instruments for medical work. Previously they had used modulated sound waves which evoked subtle radiations matching radiations emanating from the patient's cell-groups. It now struck them that they might be able to use modulated light waves to even better purpose. This idea was a correct one. A long series of experiments and a steady chain of development work led to the invention of the Colorscope which will be described later.

At the same time there was a gradual development of broadcast treatments. As mentioned earlier, it was found possible to use a patient's blood specimen for making contact with him and discovering the nature of his maladies with the diagnostic instrument. Experiments showed that it was also possible to use his blood specimen to project the correct treatment at him and restore the harmony of his vibrations. But in this case of course energies were involved which unlike acoustic vibrations took little account of distance.

Most of his earlier attempts to make broadcast treatment sets strike de la Warr now as being rather crude. One of them was an instrument made of copper wire wound round copper tintacks. The only way to find out whether these sets worked was to try them out on a patient. Like the early diagnostic sets, they proved to be very capricious because they needed a stabiliser. After he had incorporated his magnetic tuning device he wondered how his first sets could have worked more than perhaps once in the day, and then only by chance.

As time went by, most of his treatment instruments for near and distant use began to sell steadily and the reports which came in from the purchasers showed that his work was correctly based. But it became obvious that a great deal depended upon the operator. Operators had to be coached individually, and this took a great deal of time and trouble on the part of Mrs de la Warr. No instrument was sold unless the user had first been trained.

Some prospective purchasers lived abroad and had to come to England if they wished to have an instrument. Not everyone, it was found, made a good operator, but people had to be taken as they came. The de la Warrs always preferred to choose trainees from among medical students, practitioners, or recognised medical auxiliaries, for this at least ensured a basic knowledge on which to build. But even so, the number of good, fully trained operators is still not very large.

One of their best pupils was a blind American doctor for whom they made a special diagnostic instrument which clicked musically when the dials were turned so that the settings could be made by ear.

CHAPTER SEVEN

THE SPIRAL STAIRWAY OF EVOLUTION

> The whole outward, visible world with all its being is a signature, or figure of the inward spiritual world; whatever is internally, and however its operation is, so likewise it has its character externally; like as the spirit of each creature sets forth and manifests the internal form of its birth by its body, so does the Eternal Being also.
>
> Jacob Boehme in *The Signature of All Things*

As we have seen, the magnetic tuning device in the diagnostic and treatment sets obviated the uncertainty which caused the earlier instruments to work so capriciously. How was it able to do this?

The bar-magnet which is the chief part of the device stands on end between the wells carrying the blood specimens on the diagnostic instrument. When in this position the space around it is a network of stresses and strains due to its magnetism. That is to say, there is a magnetic force field in its neighbourhood. For correct tuning, the magnet with its force field has to be rotated until, for some reason, the radiations which the instrument is emitting are strengthened.

De la Warr came to the conclusion, then, that there must be an interaction between the energy of the magnetic force field and the energy of these radiations. This needed investigation, but it was not until 1947 that he was able to carry out the basic experiments.

Considering the results which these experiments achieved, his apparatus was absurdly simple. Besides the portable detector, he used for his first experiment a bar-magnet, a sheet of paper and an antenna wire which is merely a thin piece of wire to connect the detector to various points on the paper sheet.

The detector functioned by matching radiations of different kinds. If, then, there were harmony between these radiations and the magnetic field, it ought to show in what position after rotating it the magnet would give the strongest reaction.

This proved to be correct. Placing the north pole of the magnet flush with the paper and exploring with the free end of the antenna wire, the strongest reaction was found when the magnet was twisted round so that it stood in a definite rotational position in relation to the Earth's magnetic field.

It seemed that when it was in this position the magnet was radiating energy of some kind—probably, de la Warr thought, corresponding to the inner structure of the atoms or elements contained in it. If this were so, there should also be an interaction between the magnet's radiation and that of each of the elements—if the elements really did emit radiations.

Samples of iodine, copper, gold, silver, sulphur and carbon were assembled and another experiment was carried out. An iodine crystal was placed on the end, the North pole, of the upright magnet flush with the paper. With the top of the magnet as centre, a circle of nine inches radius was drawn, and then Mrs de la Warr concentrating her mind on iodine explored round the circle with the needle of the detector. She got a reaction, or resonance, at a point about thirty degrees east of north.

It looked as if the iodine were radiating a beam in a definite direction.

The iodine crystal was removed and again she explored with the detector, this time focusing her thought on iodine; and again a reaction was found in the same place. There was no doubt that the controlled thought of iodine had the same effect as the radiation of the iodine crystal, for the experiment was repeated many times, always with the same result.

The samples of copper, gold, silver and so on were now tested in a similar way, and they too produced reactions, but each in a different direction which was specific for each element. So by means of repeatable experiments it had been shown that the various elements radiated directional beams, or Fundamental Rays as they came to be called, like tiny, invisible searchlights. But the presence of the bar-magnet rotated to a certain position was needed to make this apparent, for when the magnet was rotated 'off-tune' so that it was no longer oriented, the phenomenon disappeared.

Two facts had been established. There was an interaction between the radiations evoked by air waves and the magnetic force field. Secondly there was something more to a magnetic field than the thing which every schoolboy learns in the well-known class-room experiment with iron filings; for when a bar-magnet standing upright was twisted round to a certain position it interacted with the radiations of atoms or elements, and possibly with the Earth's magnetic field.

De la Warr was particularly struck by the fact that this radiation from iodine and the other elements could be detected by using controlled thought when no sample of the element was present. This seemed to place it outside the energies listed in the text-books.

He also felt that it was very significant that a bar-magnet could be used as a mechanical agent for producing this radiative energy, or at any rate strengthening it sufficiently to make it detectable. Might this be due to the internal construction of the atoms in the magnet? The opposite ends of a bar-magnet of course attract and repel. The strength of the magnetic forces is greatest at the poles and weakest at the midway point between them. There is a pulling apart between the poles. Could there, then, be a *double vortex of energy* due to the structure of the atoms in the iron which produced these effects?

This flash of inspiration came to him in January 1948, and as time went on the conviction grew strong that it was the correct solution of the problem. The vortex of energy, he thought, should be something like a double spiral, or two spirals, one upside down, with their apices joined.

Experiments with the bar-magnet and the detector had suggested that the elements emitted Fundamental Rays which boxed the compass like the spokes of a wheel. Perhaps the elements might fit into one of the turns of a spiral?

It was quite possible that the vortex of energy might be multi-dimensional, but a spiral in three dimensions would be a better representation of it than the two-dimensional circle drawn on a sheet of paper. So the newly-appointed physicist made a solid spiral as in Fig. 7.

Once again various elements were tested with the detector, and it was found that they did produce reactions at particular points on the spiral. These points all lay near the bottom end where the single spiral was narrowest.

But atoms or elements combine to form molecules, molecules combine to form living cells; cells in their turn are the building bricks for living creatures which go to compose the integrated entity which we call Organic Life, and so on. In the same way, the Fundamental Rays of atoms, molecules, cells and so forth should each combine to build the Fundamental Ray of a bigger and more complex organism. Everything from an atom to a Universe should have its Fundamental Ray.

With this idea in mind, the de la Warrs tested with the detector things like chemical compounds, living tissues, bacteria, flowers, and human beings, and they found that each one of them fitted in at some point on the spiral.

It gradually became clear that their spiral was a three-dimensional representation or projection of a multi-dimensional magnetic spectrum. Everything had its particular place on the spiral. The simplest things, the elements, were nearer the spiral's neck, and the more highly organised things higher up on the larger turns. The loops represented levels of organisation and the spiral was a spiral stairway of Evolution.

So evidently there was an actual relationship between all things in space and time, as Professor Einstein maintained. But what were these Fundamental Rays? Surely the tapping of the spiral at a particular point represented that person's (or thing's) total experience to date. In other words, the Fundamental Ray was the end-result of all one's experiences to the present time. De la Warr was inclined to think that the concept of a spiral was not as good as it might be and that possibly the more accurate idea was the vortex of energy he had originally conceived.

The next question which arose was, could these Fundamental Rays be detected by some other method more direct than searching for reactions with the portable detector? Many forms of radiation affect photographic emulsions and it seemed reasonable to ask whether these Fundamental Rays could be photographed.

Another experiment was staged, this time with a large sweet tin in which a bar-magnet was set in an upright position. On the magnet was fixed a tube which could rotate around it, and also slide up and down the top half of it.

The object of the experiment was to see if Fundamental Rays

could be photographed as they left the magnet when 'boosted' in some way. The tube was connected up to the mains, which gave it an alternating potential with respect to the tin of 230 volts at a frequency of fifty cycles, and a small heated coil placed inside the tube.

The next thing was to determine the position of the tube corresponding to this same frequency, and this position was marked on the outside of the tin. A circular hole about the size of a penny was cut in the tin at a point facing the end of the tube.

Everything was now ready for the experiment. In the dark-room a standard photographic plate was exposed opposite the hole in the tin. When the plate was developed there was a very dark circle on it about the size of a penny. But when the tin and the magnet were rotated forty degrees so as to bring the accelerator off the beam, the plate remained blank.

The de la Warrs were as excited as the Curies must have been when they discovered the radiations from radium. At first they hardly dared trust the evidence of the plate; but they repeated the experiment several times, always with the same result.

This experiment was remarkable, for it showed that a hitherto unsuspected behaviour of energy particles could be ascertained simply by tuning in to 'the magnetic spectrum'. If the same voltage and frequency had been connected to the accelerator without the rest of the apparatus, such results would have been impossible. The introduction of the magnetic field had had the effect of satisfying certain conditions which were necessary to enable the energy to manifest.

The photographic approach to these Fundamental Rays appeared to be a much better one than the approach through air waves. A proper dark-room was needed now, and so in a few months a third laboratory was built with a dark-room attached.

On the hypothesis that everything both animate and inanimate had its correct angular relationship to the Earth's magnetic field, further experiments were carried out. Here is a brief description of them.

Experiment 1*a*

A small phial of iodine crystals was placed on a photographic plate in the dark-room, and then it was rotated until a reaction was picked up by the detector. A half-minute exposure gave the result shown in Fig. 8.

Experiment 1b

The phial of iodine crystals was rotated through an angle of forty degrees so that it was off-tune with the Earth's magnetic field. Fig. 9 shows that the beam of radiation is missing.

Experiment 2a

A small, sealed reliquary attested as having belonged to St. Ignatius replaced the iodine crystal. This produced the photograph shown in Fig. 10.

Experiment 2b

The reliquary was rotated forty degrees off-tune, and an exposure produced a nil result as in Fig. 11.

One of the most curious effects in these photographs is the clear-cut shadow of the reliquary. Such a shadow could have been produced by placing the reliquary on a piece of white paper and shining a light on it obliquely from above. But the shadow and the strong outline appeared from an experiment in which there was no visible light at all. At the position of correct tuning by rotation, the reliquary had evidently been illuminated by some source of energy. Could it be absorbing vibrations and re-radiating them?

A close examination of the plate showed that although the reliquary was in light relief, there was a certain amount of detail visible, which supported the idea.

The reliquary was now placed under the plate, and this resulted in the photograph shown in Fig. 12. Energy passing to the reliquary through the plate had this time blackened the emulsion, but only over the area of the reliquary itself. The positive and negative effect on the two plates was particularly interesting as it suggested polarity.

But what was the significance of the tuning? How was it that only when an object was rotated so that its axis lay at a certain angle to the Earth's magnetic north-south axis it became a resonator itself? It seemed that it became a resonator in virtue of some kind of radiation along its own Fundamental Ray.

Could this be the pathway and the means whereby Cosmic or Solar energy, the reservoir which maintained life, reached each living thing? Was this the secret source of life in the cell? It is true that the reliquary and the iodine crystal were 'inanimate', but was there no life even in an atom?

The magnetic spectrum in which there was a place for everything from an atom upwards seemed to imply that there was no such barrier between animate and inanimate as physical science supposed. The Fundamental Ray, or 'signature' of an entity evidently gave it a stamp of individuality, a place in the general order of existence.

Perhaps even a musical chord had its Fundamental Ray and therefore a right to exist in the universal hierarchy? Experiments were made with different chords struck on the piano, and each one of them gave a reaction at a different point in the spiral when tested with the detector.

It seemed that there was relativity of matter, relativity of existence, and that all things were related by their inner content.

It also appeared that these phenomena of life were closely connected to that of charged particles, for when some of the photographs described above were examined under a microscope, there were visible the tracks of atomic particles in the emulsion, marks where particles had collided, and so on, as is commonly seen in photographs taken of atomic happenings inside a Wilson Cloud Chamber.

By the early part of 1949, some 110 photographs had been taken in the course of about twenty different experiments. These photographs made it quite clear that the Fundamental Rays of elements might possibly consist of a particulate radiation and they also threw further light on the magnetic spectrum.

There was something ludicrous in the thought that all this had been done with a sweet tin, a piece of brass tubing and a magnet, when millions of pounds were being spent on betatrons, synchrotons, and other elaborate pieces of apparatus.

This enormously powerful and expensive atom-splitting machinery was in line with orthodox scientific thought. It produced, or was about to produce, a further expansion of the industrial age with all its complications and doubtful benefits to

FIG. 7. Approximate proportions of Spiral referred to

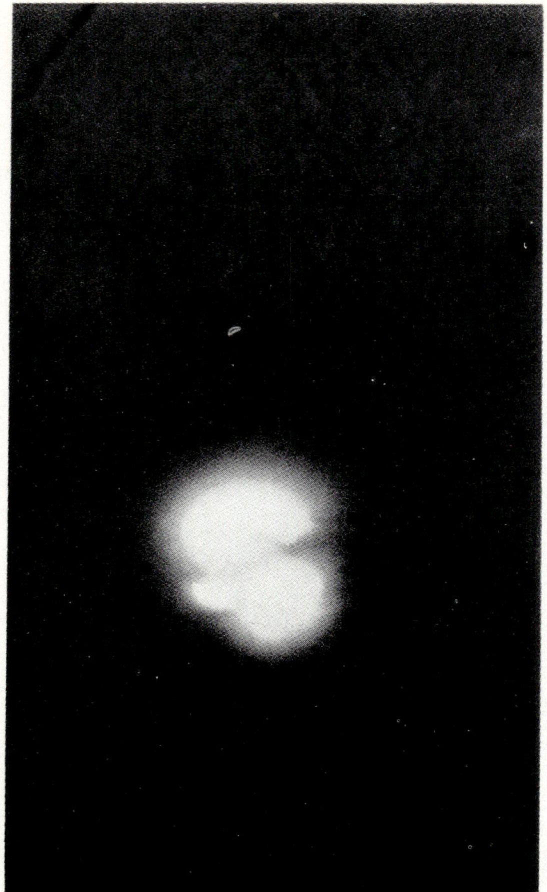

FIG. 9. Radiation from Iodine Crystals, 'off tune'

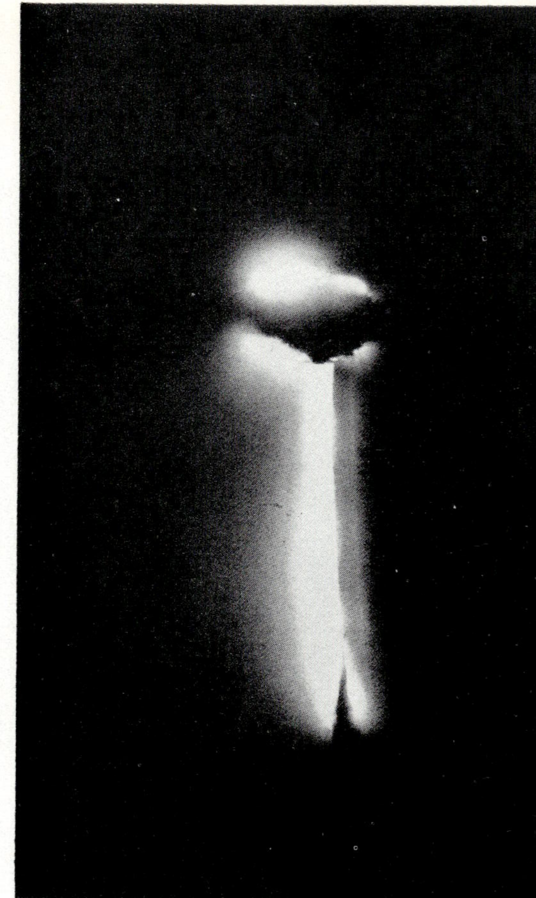

FIG. 8. Radiation from Iodine Crystals, 'on tune'

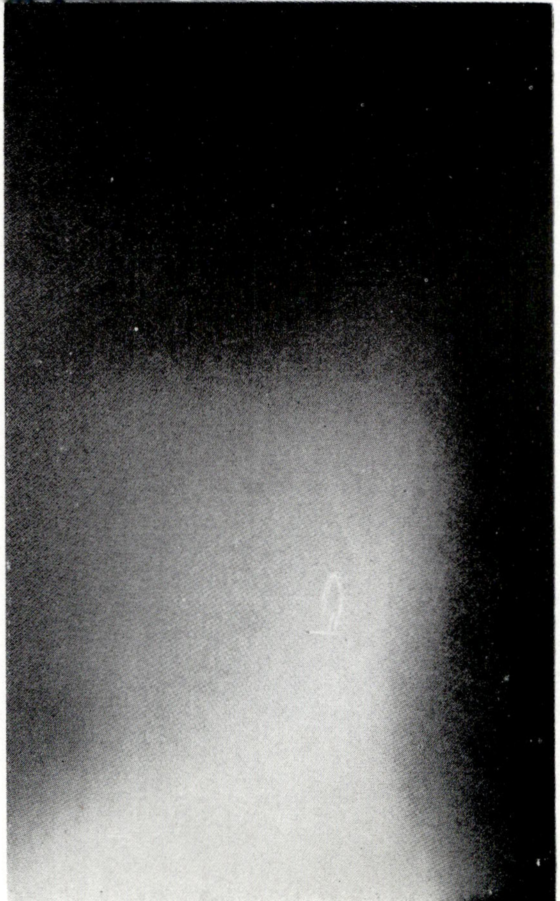

FIG. 11. Case containing relic of Saint, 'off tune'

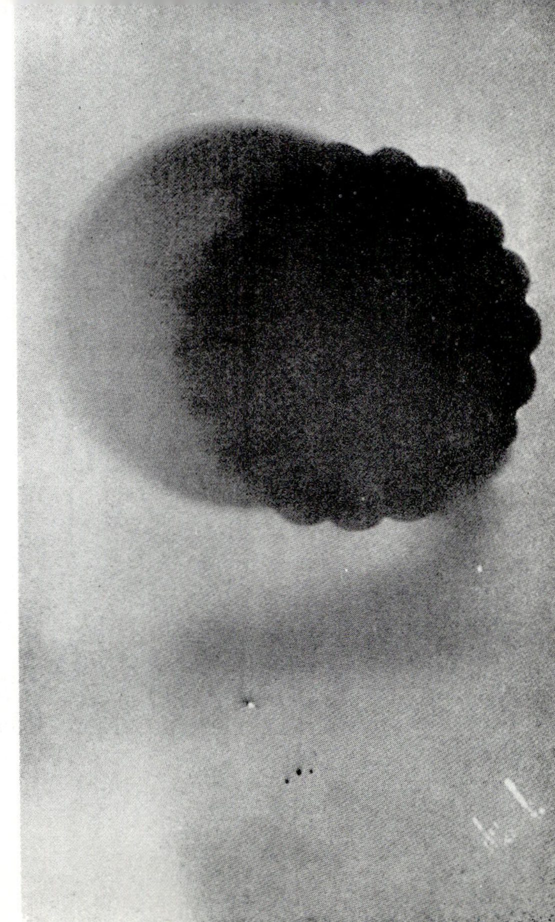

FIG. 10. Case containing relic of Saint, 'on tune'

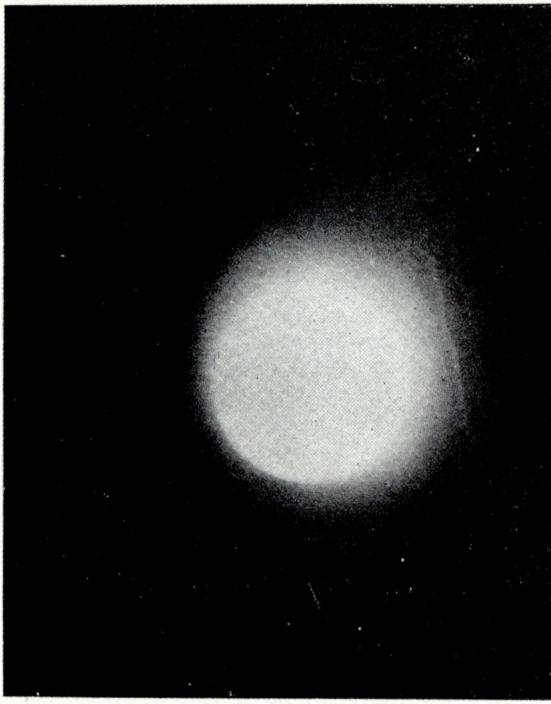

FIG. 12. Repeat of Fig. 10, but with photographic plate above relic

mankind, not to mention weapons of war which might well mean the destruction of civilised existence.

But the line which de la Warr and his little band were following was quite different in character. It had not the inherent spirit of violence, and the results which it was likely to achieve were in a different category.

This being so, the means which would lead to further discoveries were unlikely to be the same as those used by electrical technologists or atomic physicists. De la Warr had begun by accepting the tenets of the electro-magnetic theory and by trying to develop electronic instruments. But now he decided that these kinds of instruments would be useless to him and even misleading.

He had entered a new world of subtle radiations and resonances where original apparatus was called for. It was a world nearer to the heart of Nature.

CHAPTER EIGHT

USING LIGHT TO CARRY WAVE-FORMS

> For instance, a corresponding coloured ray directed upon any active element of opium transformed it into another active element which corresponded in its newly-acquired vibrations to the vibrations of the coloured ray which had acted on the given element. . . . The same result was obtained if instead of these, coloured rays, corresponding sound-vibrations . . . were directed upon this same active element.
>
> G. I. GURDJIEFF, in *All and Everything*

DURING six years of research work, attempts had been made to radiate treatment to patients who might be miles away through the medium of their blood specimens. This technique seemed capable of further development, and so the next step was to incorporate a magnetic aerial which would enable the operator to tune his treatment instrument to a state of resonance with the patient.

With such tuning it was found that the results were good. It meant a great deal of additional work, since the aerial had to be tuned each time a new patient was put on the list for treatment, but by this time Mrs de la Warr had organised a routine for continuous broadcast therapy.

The normal method is to take a blood specimen of the patient on a piece of white blotting paper and to use this to link him with the operator. The blood is not chemically analysed by the instrument, but it provides the necessary link and seems to emit something like a carrier wave. At the same time we must not forget that the operator is using an instrument which functions in a region beyond the physical (in its more material sense) and that he is assisted by a form of energy which is subservient to the power of his mind.

Using this instrument, he is able to tune-in to his patient's vibrational rate, discover what diseases or infirmities he is suffering from, and also the severity and location of these maladies. He can then place the blood specimen in the well of his treatment set and radiate suitable vibrations to the patient in order to neutralise the vibrations of the disease from which he is suffering.

Some actual cases will perhaps make the matter clearer.

A lady of fifty-six came to Mrs de la Warr suffering from a spasm which gave a recurrent head movement. She had been treated for many months by some of the best orthodox practitioners, but the spasm grew steadily worse until at length a series of atropin injections almost paralysed her and she refused further medical aid.

She was found to react most strongly to the rate for spasmodic torticollis, and so with the first four dials set at 60, 4, 6 and 2, the severity of the disease was measured on the last dial. This proved to be ninety per cent.

With the help of the Location and Detail Sheets placed under the sliding cursor, Mrs de la Warr next discovered what parts of the organism were contributing to the torticollis and therefore vibrating in sympathy with this, the torticollis rate. Three locations were found:

The Muscular System	Sheath of the Trapezius Muscle
The Nervous System	Posterior Cervical Plexus and Sub-trapezial Plexus
The Skeletal System	The Vertebrae

The next thing to discover was what was wrong in these locations. Using the Cause Sheet she found that:

The sheath of the trapezius muscle showed injury and the presence of *Mycobacterium Tuberculosis.*

Both the posterior cervical plexus and the sub-trapezial plexus showed vitamin deficiency, mineral imbalance and the presence of both *Mycobacterial Tuberculosis* and *Clostridium Tetani.* One of the cervical vertebrae recorded a fracture.

It looked as if the *Clostridium Tetani* were causing a contraction of the trapezius muscle by irritating the motor nerve. As to the fractured cervical, the patient now admitted that while travelling by train some years earlier, a suitcase had fallen on her head. X-ray examination confirmed the fracture.

Treatment now began. The broadcast treatment rates were:

6088	T.B. in Trapezius Muscle
1092	Injury to Cervical Vertebra
6088	T.B. in Sub-trapezial Plexus
7096	*Clostrid. Tet.* in Sub-trapezial Plexus
2544	Vitamin B for Post. Cervical Plexus
3803	Gold for Post. Cervical Plexus.

At intervals the severity of each malady was measured and found to be steadily diminishing. After several months of treatment the spasmodic movement lessened, and in time the patient fully recovered. She is willing to corroborate these facts.

Another interesting case was that of a lady of over eighty who was living 200 miles away in Cornwall, bent double with rheumatism and in much pain. At this distance she was treated for a succession of different conditions: toxins, hormone imbalances, mineral and vitamin deficiencies, affection of the blood, muscles and bone marrow. By means of her blood specimen she was checked up from day to day in the Laboratories at Oxford, and each day the disorder which was most marked was treated by radiating the appropriate wave-form.

She straightened up in a most remarkable manner and after a year she became once again active, travelling alone on buses to do her shopping, taking walks, doing her housework, and in fact behaving very differently from the helpless woman she was rapidly becoming before her first treatment. Later, she moved to Oxford where she received regular treatment by personal irradiation. She is now being treated by her daughter who has made a close study of de la Warr's work. For a woman of her age her condition is excellent and she is willing to corroborate these facts.

Another case of a different kind was a lady in her middle thirties who could not bear the thought of travelling. She was at once sick if a bus ride was suggested. Mrs de la Warr found a reaction to the rate for lesion in the cerebellar cortex, inflammation in the myocardium, and tissue derangement in the ventricle walls of the heart. There was also trouble from *B. Coli* in the kidneys.

Treatment consisted in normalising the cerebellar cortex and the pituitary gland, and inhibiting the growth of *B. Coli.* The heart condition responded well and became normal when the other troubles were treated. In three months the patient was able to take long journeys without any suspicion of her previous symptoms. After going abroad by air, she made a special journey from Scotland to see Mrs de la Warr whom she had never met. Her treatment had been in the form of radiation broadcast from Oxford. She is willing to corroborate this.

In scientific research work the knowledge gained is often empirical, at least to begin with. Pioneers may find that something works without knowing why it works. Like others, de la Warr had found that a blood specimen was permanently in resonance with the body of origin, but evidently there was much to be learned about the sort of radiation it emitted, why rotating it should affect the strength of this radiation, and so on. Perhaps photography would throw some light on these questions as it had on Fundamental Rays and the magnetic spectrum.

De la Warr placed a blood specimen absorbed by a small piece of blotting paper on a photographic plate and rotated it at thirty-degree intervals making a thirty-second exposure in each position. This entailed twelve exposures, but the results were not very striking. True, there was a noticeable difference in the intensity of the radiations from the blood in these different positions but it was not quite what he had expected or hoped for.

Something was needed to evoke the energy-pattern. In the next experiment he used the blood specimen of a patient who was suffering from rheumatism. Placing it at a critical distance from a treatment instrument with the dials set for rheumatism, he rotated the specimen to such a position as to give the strongest reaction when using the portable detector. He then switched on the treatment instrument and irradiated the blood specimen.

All these things should surely combine to produce something significant on the photographic plate. They did, as may be seen from the spectacular result shown in Fig. 13. Something of this kind had been expected in the previous photographs, but evidently certain conditions were needed to bring it out. These conditions had now presumably been satisfied, but a great deal remained to be clarified and explained.

It was found that when the experiment was repeated with the blood specimen rotated say forty degrees so that it was off-tune, no photograph could be obtained even when it was being irradiated with the correct wave-form for rheumatism. The explanation of all this seemed to be that *there was a force field round the blood specimen when the latter was correctly oriented.*

What was the significance of this force field? What did it express?

Possibly it expressed the radiation pattern of the subject which had some relation to his state of health. In the case they were experimenting with, the patient was suffering from rheumatism. The force field of his blood specimen would therefore include the wave-form of rheumatism; hence it could be used for the detection of rheumatism and also for tuning-in when it was wished to apply the corrective treatment.

You will remember that his earlier experiments with the light-valve had led de la Warr to the notion that light waves might suit his purpose even better than acoustic waves for carrying or evoking the subtle energies which were needed for treating a patient. He conceived the idea of using a variety of coloured beams. Each beam, being of a specific frequency, would perhaps act as a carrier for a different wave-form of energy.

How could such an instrument be designed? Sitting one afternoon in the Cumberland Hotel, London, discussing this problem with a member of his staff, they had a flash of inspiration. They pictured an instrument consisting of eight revolving turrets, each of which contained seven different colour filters and one blank. All would be arranged in a line so that any colour or combination of colours could be produced by turning knobs on the front face of the apparatus.

By turning another set of knobs on the back face, treatment wave-forms could be created, each to be combined with or carried by the appropriate light-beam. They rapidly designed an optical system, and in the short space of three weeks the Colorscope was working.

When it was tested it proved to be very effective, but it was by no means plain sailing for the de la Warrs. Not being qualified doctors, they were prevented by Law from treating cancer, tuberculosis, diabetes and other grave maladies except under the supervision of a medical man. The few patients who were sent to them by practitioners were usually difficult advanced cases with very little hope of recovery.

So there was little chance of discovering what the Colorscope could really do. Nevertheless the results they achieved were most encouraging. Few cases are ever simple and straightforward. For instance, mere inhibition of the multiplication by fission is not enough in a case of T.B. There are tissue damages to be repaired, mineral deficiencies to be restored, glandular imbalance to be rectified, and so on. But it was felt that such things might possibly be done by suitable wave-forms carried on a colour-beam to the patient.

Ever since it made its appearance, the Colorscope has caused some confusion of thought among people who imagine that it is the colours themselves which have the healing virtues. Although de la Warr agrees that various colours may produce resonance with certain cell-groups in the body, with organs, blood constituents, bacteria, and so on, he thinks this can have very little therapeutic value by itself. The all important thing is the specific wave-form of vital energy which accompanies the light waves.

At first it appeared to de la Warr that this wave-form was *carried* by the light waves, but ultimately he formed the opinion that it was not a true conception of what really happens. It seemed more probable that an omnipresent Cosmic Energy existed which can always be evoked under certain necessary conditions.

His experiments showed that it can be evoked by light waves, radio waves, acoustic waves, or even possibly by a factor that was common to all these sources of energy. So in treating a distant patient it is not strictly speaking an emission of some kind of energy which has to traverse the intervening distance, but an evocation of Cosmic Energy which manifests in the physical organism of the patient in a pre-determined form. It was, at all events, an unknown principle but it looked like being of great value when properly applied.

In the case of the Colorscope, the dials at the back of the instrument are set to form the requisite wave-form for treatment, and this enables Cosmic Energy to manifest in the form determined by this pattern. The process is that of evoking or channelling Cosmic Energy to, say, the heart muscle of the patient in a form exactly similar to the energy which the heart muscle would receive naturally if the patient were in a normal state of health.

Where disease is concerned, it is easy to imagine the reverse process of channelling. If bacteria are tiny radio stations broadcasting their special programmes, then the Colorscope has the effect of jamming these programmes.

This being so, it looked at first as if the efficacy of the instrument could very easily be demonstrated. All that had to be done was to irradiate bacterial cultures with the Colorscope, and any pathologist would be able to see at once that it worked. So obtaining some cultures of Staphylococcus prepared at a neighbouring hospital, he and an assistant sat up night after night into the small hours of the morning trying to destroy these bacteria; but nothing happened. What was the reason for this unexpected failure?

The answer which gradually dawned on him threw an extraordinary light on the world into which he had penetrated.

Scientists regard energy as impersonal, or completely lacking in qualities such as beneficence or maleficence. But this Cosmic Energy evidently had a beneficent quality. As far as could be judged, it maintained life, and it even healed and restored. In fact, it appeared to be related in some way to the source of life. So if the bacteria of some disease exist in a patient, it will prevent them from multiplying or will even destroy them in the process of healing the sufferer. But if the bacteria are in a culture outside the patient it will not kill them because they are living creatures existing in their own right.

In one respect this was a disappointing discovery since it meant that the normal routine of the pathologist could not be used to prove de la Warr's case. However, the Colorscope soon proved itself a very effective instrument. A couple of early cases will show how well it works in the hands of a capable operator.

A more or less bed-ridden Colonel who received treatment for rheumatism wrote:

> The Colorscope has been most successful and has been of

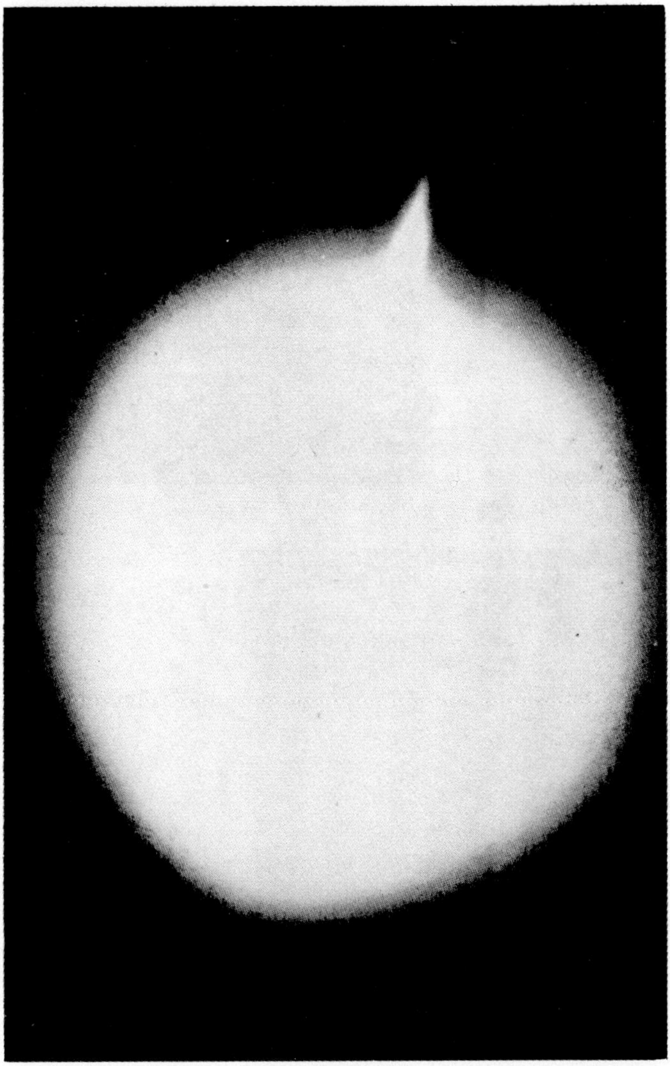

FIG. 13. Radiation from blood specimen of rheumatic patient

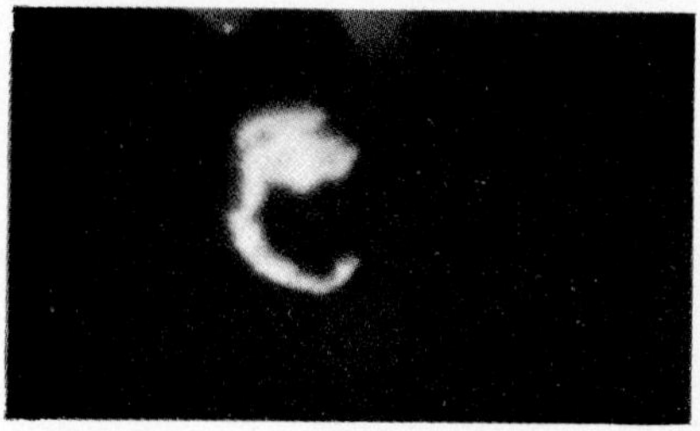

FIG. 14. Malformed foetus fourteen days old

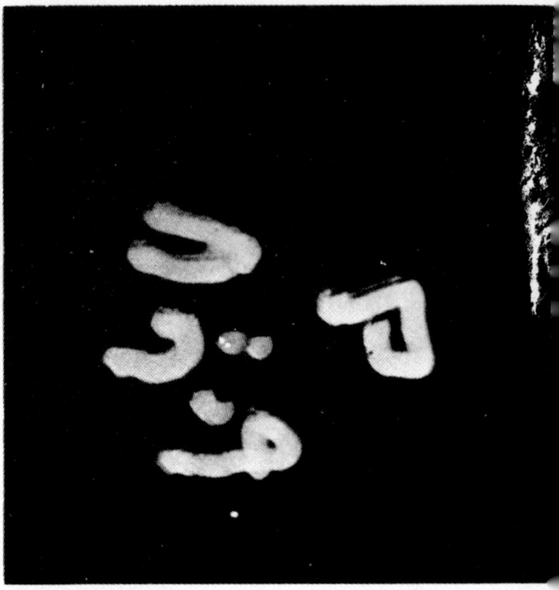

FIG. 15. Cat pregnancy

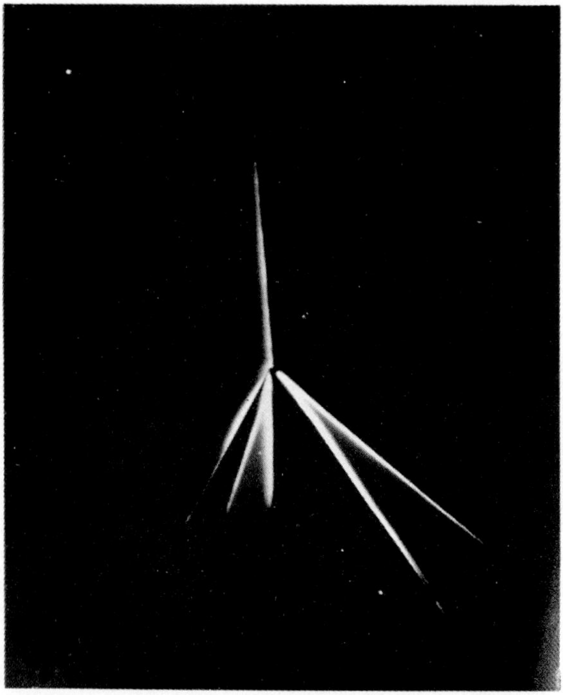

FIG. 16. Radiations from crystal of copper sulphate showing elements

inestimable value to me. It has certainly not removed my rheumatoid arthritis, but it greatly reduced it and has certainly halted its activities. I am glad to say I can do considerably more than I used to be able to do—in fact, I can now do most things I want to. I believe that if it had not been for the Colorscope I should have been more or less crippled by now.

Another patient with a cyst in the throat showing signs of potential cancer had Colorscope treatment for six months. The cyst responded well and was dissipated completely. There has been no recurrence after two years, but it will be necessary to wait four or five years before a complete cure can be claimed. However, treatment has dissipated the obstruction in the throat and the patient is able to lead a normal life. He is willing to corroborate this.

As in all such cases, the treatment varied from time to time according to which cell-groups of the throat were needing special attention. Since it is possible to tune-in to the various constituents of the blood separately and assess their contribution to the Harmonious Whole, the condition of the blood could also be corrected as and when might be necessary.

It is interesting to compare these methods of treatment for a disease such as cancer with the orthodox treatment by deep X-rays. Deep X-ray treatment destroys healthy cells along with diseased cells. Such a method is hardly in line with other less crude methods used in modern medicine.

On the other hand, the new methods described in this book are far more delicate and discriminatory. Unlike a general massacre of friend and foe, they are equivalent to a resistance movement which arms and fortifies the inhabitants against the occupying power, supplying them with guns and food and at the same time engaging the enemy by highly selective bombing.

CHAPTER NINE

DISEASES PHOTOGRAPHED THROUGH BLOOD SPECIMENS

> The knowledge of Nature is the foundation of the science of medicine. If you wish to be a true physician you must be able to do your own thinking and not merely employ the thoughts of others. To be an alchemist is to understand the chemistry of life. Medicine is not merely a science, but an art; it does not consist in compounding pills and plasters and drugs of all kinds, but it deals with the processes of life, which must be understood before they can be guided.
>
> PARACELSUS

A MOMENT of exaltation which de la Warr will never forget is the moment in which he realised that he could diagnose a disease in a patient by photographing his blood specimen. It was a moment such as comes to a traveller when he catches the first glimpse of his destination. Whatever the obstacles, he knew now that one day his work would be acknowledged and applied.

The discovery came about in this way. Since he had found it possible to take the remarkable photograph shown in Fig. 13 by irradiating the blood specimen of a rheumatic patient with the wave-form of rheumatism, he felt sure that it would be possible to detect diseases by means of the blood provided he could invent some suitable photographic method.

The method which occurred to him was to lay several blood specimens on the same plate and then irradiate them with the wave-forms of the diseases from which each of the patients was suffering. If this were done, the effect should be to make visible the radiation of each of the blood specimens in turn, because only the blood which was being treated with its corresponding disease-vibration ought to affect the emulsion.

Three blood specimens were chosen for the experiment, the owners of which were suffering from tuberculosis, diabetes and nervous disorder. When they were irradiated with the wave-form of tuberculosis, there was a dark area on the plate under the tuberculosis and also the diabetes specimens—because as it happened, the diabetic was also developing T.B.—but none at all under the third one.

This suggested that it might be possible to develop a technique for the visual confirmation of a diagnosis, and it was also a strong link in the chain of evidence supporting the peculiar fact that controlled thought is an operative factor.

The chain of evidence was this. The table of rates for diseases had been mapped out by a method which orthodox science would consider fantastic, namely by relying on the diagnostic instrument which operated to a great extent on the mental plane.

In order to use the diagnostic instrument successfully, the operator has to think of the patient's condition which he is trying to detect. For at least one second he must hold in his mind a single thought—as a result of which his solar plexus emits a wave-form which resonates with the blood specimen. This is a prime essential for the functioning of the detective device.

No suggestion or imagination is involved. Suppose he passes his fingers over the rubber diaphragm with the blood specimen of a cancer patient in the well and with the dials set for the cancer rate. If his mind is blank or wandering, there will be little or no sign of a 'stick', but as soon as he concentrates mentally on 'cancer' he will feel the rubber spring to life and cling to his fingers, indicating resonance.

And now the table of rates which had been arrived at by the use of controlled thought had been proved correct by photography. It was clearly so, because if these rates had been wrong, no such distinctive results could have been obtained.

Here, then, was photographic proof of a direct relationship between mind and matter. It showed beyond doubt that whenever we think clearly of a person or an object, we are in resonance with it. By thought we can link ourselves to people and things and establish a two-way traffic of vibrations or influences.

When in 1950 de la Warr had established the beginnings of his new technique for diagnosing photographically, he felt that his feet were on firm ground, but he knew that this technique must be developed and improved before it could become a practical proposition. A patient might be suffering from any one among hundreds of possible diseases, and to arrive at a correct diagnosis by trial and error would be a most laborious process.

The Camera does not analyse a blood specimen, but uses it to set up a condition of resonance with the patient. It is something like the automatic telephone system which connects a subscriber when he dials a number. The same thing can be done by using other things from the subject's organism, such as a sputum slide, sweat, hair, or a piece of skin. These 'witnesses', as they are called, have force fields which are in permanent resonance with the force field of the person to whom they belong, but resonance is strongest in the case of blood or sputum.

Force fields, it will be remembered, are something like a close network of radiations from an immense number of tiny aerials all broadcasting at once. Was it possible to sort out these radiations by selective tuning and pick up the various 'programmes' separately?

The apparatus which de la Warr had already used was obviously capable of being selectively tuned, and it had produced beams of radiation which affected the emulsion on a photographic plate. A lens system was devised to focus the radiation more accurately onto the plate, but something else was needed if the complexity of the radiations was to be sorted out.

At this point in their work the de la Warrs had what was probably their most crucial idea to date. It was to use two intersecting beams of radiation, one of which emanated from the patient's blood specimen while the other was a synthetic radiation specifically tuned to some particular disease or condition. This second beam would 'analyse' the first beam at the point of intersection.

When this apparatus had been fitted up, a blood specimen was chosen for the first experiment and the plate was placed at the point where the two beams crossed.

This specimen was from a woman who wished to know if she were pregnant. Tuning to the foetus rate of vibration, a photograph was taken which is shown in Fig. 14. Obviously it is of a malformed foetus. The woman was in Ireland, whereas the photograph was taken in Oxford.

There was naturally some excitement at such a spectacular result. To photograph a foetus in a distant subject by means of a blood specimen seemed sheer magic, but here was conclusive proof that it could be done. The Camera was a first-class witness and must surely carry conviction. Once accepted and applied, this technique could among other things revolutionise the entire system of ante-natal care.

Looking round for another mother-to-be on whom to test the new appliance, they remembered the Laboratory cat, Tinker. No one volunteered to stick a needle into her to get a blood specimen. Waving a piece of filter paper, which seemed to fill the animal with suspicion, one of the staff at length managed to catch her and obtain some sputum from the side of her mouth. This resulted in the photograph in Fig. 15.

Assuming that the two smaller objects are end-on to the lens of the Camera, you will see that there are four big somethings and two little ones. There was elation in the Delawarr Laboratories when four weeks later Tinker gave birth to four healthy kittens and two weaklings!

They were now in the position of travellers with an unknown world to explore. Next they tried taking photographs of the radiations from crystals. It so happened that they had been experimenting with the effects of these radiations on the rate of crystallisation in saturated solutions of copper sulphate, silver nitrate and alum, and these particular crystals being handy, they were chosen for the next experiments.

A copper sulphate crystal was placed in the Camera. They then tuned in to the wave-form of the copper sulphate molecule and obtained the photograph in Fig. 16.

The photograph is a composite picture of the directional beams of the atomic radiations in the molecule.

To make sure that there was no mistake about it, the Camera was tuned to detect each of the atomic rays in turn. Six photographs were taken, and each of them showed a ray in a different direction. Two of them are shown in Figs. 17 and 18.

Placing these six photographs on top of each other, the result exactly tallies with the composite picture in Fig. 16.

Here, then, was visual proof that each atom emits a directional beam of radiation and that when atoms combine to form molecules, these beams form a combined pattern.

The silver nitrate and the alum crystals were photographed with results as shown in Figs. 19 and 20.

Photographs were taken of a bottle of Oxford tap water and of a bottle of water from a spa, as in Figs. 21 and 22.

Notice how much richer in various elements is the spa water. This method of gauging the qualities of water supplies could obviously be used for foods and other substances, for even trace elements can be detected. But a quantitative analysis may not be possible because the length of the radiating lines is no guide in this respect.

The Camera gave still more interesting results in the biological field. More or less at random a photograph was taken of a common daisy which showed the pattern of its elements, rather as in the photographs of the crystals.

But this is not like ordinary photography: it is multi-dimensional. In other words, it is possible to photograph an object on different levels of organisation or existence. Or to put it differently, higher up or lower down on the Spiral Stairway.

The photograph of the daisy was taken on the level of its atoms because the Camera was tuned to this rate of vibration. But such things as daisies have higher levels of organisation and by suitable tuning it is possible to photograph it 'higher up'. For example, on the level of its cell-groups, or that of its existence as a complete entity.

If you examine some of these photographs you will notice what looks like a small spiral at the source of the rays. What is this? The attempt to discover what it was led to further experiments.

It had already been found that these Fundamental Rays were affected by the Earth's magnetic field so that no photograph could be taken unless the crystal, or whatever it might be, was correctly orientated. An additional magnetic field provided by a bar-magnet ought also to affect them. Some photographs were taken to find out if this was true.

When the test was made, it at once became clear that the magnet adversely affected the photograph except when the object was in one particular rotational position, when it increased the clearness of the ray. This of course agreed with previous experiments carried out with the portable detector.

In one experiment a single element, sulphur, was photographed and the sulphur beam appeared on the plate. The sulphur grains were then placed on the north pole of the bar-magnet which was rotated off-tune, and another photograph was taken. This time only a pin-point was visible: the Fundamental Ray had evidently been distorted or cut short. But when the magnet was rotated to the correct on-tune position and another exposure was made, it showed the full ray as before.

It was a remarkable fact that it was the rotational position of the magnet which affected the photograph and not merely the presence of the magnetic field. In classical atomic physics the behaviour of charged particles such as electrons or protons in a magnetic field is well known, but to find energy particles which are affected by a magnet in one position and not in another position is something new to science.

Electronics has developed into a separate and highly important branch of science. It has given us radio, radar, television and a host of new appliances for peaceful and warlike uses, as well as a certain insight into the atomic world. So much work had been done in this direction that, on first thoughts, it is hard to believe that a discovery, such as is described in the last paragraph, could possibly have been overlooked.

But this is to misunderstand the nature of modern scientific research work. The laws which govern matter and energy are so complex in their ramifications, and the phenomena of the animate and inanimate worlds so well-nigh limitless, that the tendency in research groups is always to keep within strict limits lest their efforts be dissipated.

This produces material results, but it narrows the viewpoint. How often must such workers come upon some strange fact or flash of inspiration which if followed up must surely lead to something important, but which being 'off the beam' must be left severely alone?

CHAPTER TEN

PHOTOGRAPHING A FLOWER FROM ITS SEED

All things, by immortal power,
Near or far,
Hiddenly
To each other linked are,
That thou canst not stir a flower
Without troubling of a star.

FRANCIS THOMPSON

A SERIES of photographs was taken of seeds and flowers. Fig. 23 is the photograph of an aconite flower.

At first sight there is nothing very remarkable about this photograph, but actually it is almost 'magical' because it was taken by placing in the Camera a bottle of homeopathic *Aconitum Napellus* pilules and tuning the Camera to the vibrational rate of the flower. It is thought that the photograph is the energy-pattern of the Monks-hood or aconite flower which determines the growth of the plant.

The Fundamental Ray of every living creature seems to contain an inherent pattern, or rather a progressive series of patterns, and when by an act of procreation or germination a new life-cycle begins, the fertilised ovum develops in accordance with these patterns. Early in the life-cycle there is a very strong growing force, or potentiality to fertility which in Fig. 23 is shown by the bright spots.

In Fig. 24, which is a photograph of a lily seed, the Camera has been tuned to 'potentiality to germination', and the series of bright circular spots shows clearly where this potentiality is strongest. In the second photograph, Fig. 25, the Camera was tuned to 'potentiality to flower'.

Figs. 26 and 27 show the same thing in the case of a horse chestnut. In the first photograph there seem to be two different kinds of fertile areas with a process of attraction taking place between the large and small ones.

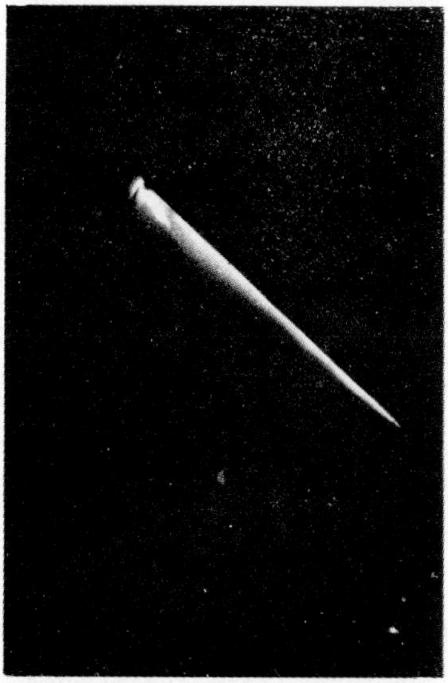

Fig. 17. Radiation from copper sulphate crystal with Camera tuned to detect Hydrogen

Fig. 18. Radiation from copper sulphate crystal with Camera tuned to detect Sulphur

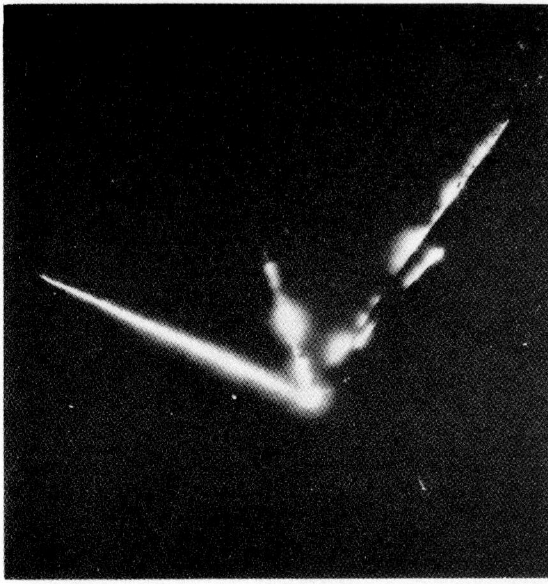

FIG. 19.
Radiations from crystal of silver nitrate showing elements

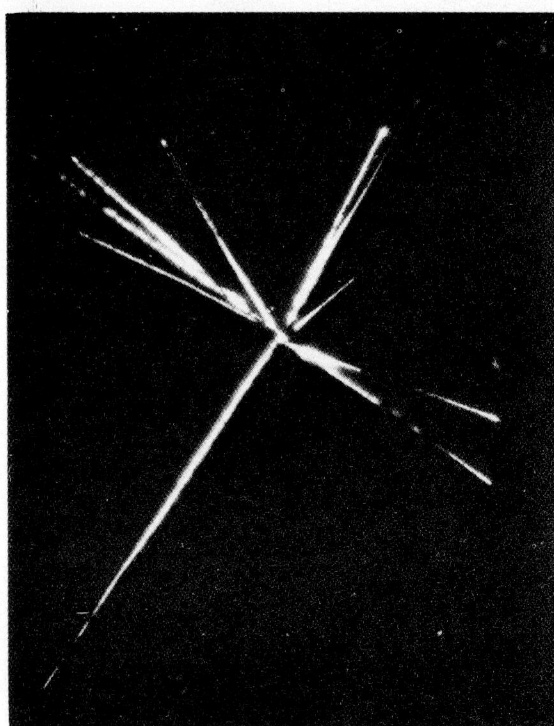

FIG. 20.
Radiations from crystal of alum showing elements

Fig. 22. Radiations from bottle of spa water

Fig. 21. Radiations from bottle of Oxford tap water, 1950

FIG. 23. Radiations from bottle of *Aconitum Napellus* pilules

Fig. 27 taken on the vibrational rate of 'potentiality to flower' gives a picture of the chestnut flower which *may* materialise. For some unexplained reason the Camera would not respond to a tuning for 'chestnut tree'. It was thought that the season of year, Spring, must have had something to do with it.

If we consider the beginning of a new life-cycle, say the procreation of a human baby, it begins on the cellular level as a fertilised ovum, and passes through progressive stages until it becomes an adult human being. At any time in the course of the baby's development the Camera will photograph it as a whole entity on the level it has reached. The rate of its vibration steadily changes and the Camera can be tuned-in to whichever rate is desired.

Suppose, then, a woman who is pregnant wishes to know when her pregnancy began, a series of photographs of the foetus can be taken on different rates of vibration corresponding, say, to one month, three months, or six months, and the tuning which gives the clearest impression on the plate will indicate the time-period.

This can be plainly seen in Figs. 28 to 31, which are photographs of the same foetus all taken on the same day, but on different rates of vibration.

The vibrational rate for three months, in Fig. 30, gave the best result because the foetus was in fact about three months old. Fig. 31, taken on a six months' rate of vibration, looks almost like a threat of a miscarriage, but this is only because the embryo was not yet emitting the radiations corresponding to a six months' growth. On the other hand it had already passed the stages of growth corresponding to the dial settings in Figs. 28 and 29.

At the time when these photographs were taken, de la Warr had not yet thought of using his Camera for sex determination, but since then he has been developing a technique of tuning to the vibrational rate for ovaries in the case of a female foetus, and testes in that of a male, and these two rates give very definite photographs at a fairly early stage.

In exactly the same way as in the growth of an embryo, the change in the vibrational rate during the progress of a disease will indicate the stage which the disease has reached.

Diseases appear to begin in the force field body as 'potentiality to cancer', 'potentiality to tuberculosis', and so on, so that the tendency to develop a disease can be ascertained before any physical symptoms have appeared.

As the disease develops, the tuning alters, each tuning representing a definite stage in its progression. This shows what accurate results can be obtained from precise tuning of the dials. Nevertheless, the Camera is not strictly speaking a diagnostic instrument: it will only confirm or refute a diagnosis. To use it as a diagnostic appliance would be like searching for some radio station of unknown wave-length with a highly selective set, switching off and on before each new attempt.

It soon became obvious that there were other uses outside the medical field to which the Camera could be put.

In 1950 the surprising fact was discovered that an ordinary cabinet photograph of a person would act as a 'link' for tuning-in to him, though not such a good link as a blood specimen or a sputum slide. It seemed to de la Warr that in some way the radiations of a person were transmitted to and re-radiated by the emulsion on a photographic plate.

If this were so, perhaps the radiations of inanimate objects would also be re-radiated by the emulsion, in which case a Camera could be developed for prospecting for metals and minerals? He modified his original camera and produced the Mark V model which was designed specially for this purpose.

Burying a small box of copper sulphate crystals about two feet deep in the garden, he took a photograph of the landscape from his bedroom window with an ordinary quarter-plate camera.

If his theory was correct, the emulsion on the plate now contained the radiations of copper sulphate at some particular point on it, so that it should be possible to locate the site simply from the photograph.

So the landscape photograph was placed in the Prospecting Camera which was tuned to 'copper sulphate'. When a photograph was taken, it showed a reaction to the pattern of copper sulphate.

Removing the landscape photograph and repeating the process with the Mark V Camera, the result was a blank.

The radiations contained in the emulsion were evidently in

resonance with the copper sulphate crystals which occupied a position in the garden corresponding to the position on the landscape photograph. If the crystals were removed, there should no longer be resonance, and consequently no photographic reaction in the Prospecting Camera.

The crystals were dug up and removed, the landscape photograph was replaced in front of the Mark V Camera, and another photograph was taken. It was a blank. Ten times this experiment was repeated, and each time there was the same blank result.

This seemed to show that it was possible to prospect with aerial photographs by a process of elimination. Crystals were buried in three different places and another landscape photograph was taken of the site, after which the plate was divided into a number of small segments with a diamond cutter. When these segments were photographed separately with the Mark V Camera, only those which included the sites of buried crystals showed photographic reactions; the others were blank.

A further experiment was carried out with one of these pieces which had shown a reaction. The emulsion was scraped off and a photograph was taken of the clear glass. The result was a blank. But when the scraped off emulsion was placed in the Camera, there *was* a reaction. It seemed certain, therefore, that the radiations which produced resonance with the crystals were contained in the emulsion.

De la Warr decided to build an improved Prospecting Camera and in due course there appeared the Mark VI model, which focuses energy so as to give photographs of subterranean deposits without even the necessity of taking it to the site. It may take several years to perfect, but it is worth time and trouble for prospecting offers unusual possibilities. In gold mining, for instance, the reefs are often erratic and there is bound to be a good deal of guess-work. The new technique should give far more precise and reliable information than is possible when using even the most up-to-date alternative methods.

An interesting experiment was carried out with the aid of Colonel M. who is able to detect water with a divining rod. Colonel M. was interested in locating suitable sites for boreholes near Kuwait on the Persian Gulf where drinking water has either to be distilled or brought many miles by sea.

It was arranged that he should send the Delawarr Laboratories some 35-mm. snapshots taken of the ground over possible subterranean water supplies located with the divining rod. The negatives would be placed in the Mark VI Camera and the sites tested for potable and brackish water.

When this was done with a photograph of a spot called Site 13, a photograph was obtained. To obtain it, a sample of brackish water was placed in the Camera, which showed that the water under Site 13 was brackish; for when a sample of tap water was placed in the Camera there was no reaction at all.

In due course a bore-hole was sunk at this point which struck an artesian supply of brackish water. Thus a Camera in Oxford had accurately located a subterranean water supply in Iran, and it had even gauged its quality.

Not only metals and minerals, but strayed animals and missing persons can be tracked down by these new photographic methods.

An experiment was made to see if the Camera could identify a photograph from a finger-print. Taking his secretary's fingerprint and using this as a link, de la Warr mixed up her photograph with a number of others, and on making the test he found that the Camera would only give a reaction when the correct photograph and fingerprint came up.

No doubt this method could be used instead of the cumbersome identification parade. And just as water, coal, oil and so forth can be located by prospecting, so a missing criminal could be tracked down by the police who would 'prospect' with the aid of an aerial photograph of a city, the criminal's photograph or finger-prints, and a Camera at police headquarters tuned to detect his radiations.

It can be disclosed now that by means of his photograph, Pontecorvo, the scientist who went to Russia, was located by the Mark II Camera in the outskirts of Leningrad three weeks after his escape via Italy. The correctness of this result was subsequently confirmed.

His photograph was placed in the Camera and radiated with various latitude and longitude 'fixes' until a condition of resonance was established with the part of the world where Pontecorvo was. There is a separate tuning of the apparatus for 'position in space'. Later on, this Mark II Camera was superseded by the Mark V and Mark VI models which are more efficient for prospecting.

It is only a short step from this to detecting enemy movements in war-time. Camouflage is becoming more elaborate and effective as was shown in the fighting in Korea and Indo-China. But no camouflage however elaborate would conceal the presence of troops, guns and armoured vehicles photographed from aircraft by day, or by night with infra-red rays, and then analysed by the Mark V Camera.

It is true, infra-red photography has now been developed to a very high standard so that it is able to detect even well-marked foot-prints, but it cannot pick out underground installations except inferentially from the discovery of entrances and exits.

An exceptionally useful application would be in the detection of minefields. The new photography would be much more expeditious than the laborious method of searching for mines on the ground with an electronic detector. Presumably atom bomb piles in enemy country could also be located. Indeed, when these Cameras are perfected there is no reason why almost anything could not be traced from a submarine to a foreign dictator.

In war-time, even more than in peace-time, fatigue in metals is a serious matter, especially in aircraft engines. Experiments show that a technique for testing the working parts of aircraft engines could be developed on much the same lines as that for diagnosing diseases in patients. Nor is there any reason why this should be done only when machines are grounded. Although a test has not yet been made, it is believed than an aircraft could be vetted while it is in flight, providing it maintained its course while the photograph was being taken.

Why this proviso? Because de la Warr discovered that when taking photographs of patients with his Mark I Camera any rapid rotational movement on the patient's part during the exposure produced a blurred picture.

This was only one of the surprises which the Camera had in store for him. As we shall see later, it had many more, some of which were so strange that it almost seemed as if the Camera was bewitched. Each time the Camera behaved in an unaccountable manner a new line of research work had to be undertaken in order to find out the cause of it. This was sometimes irksome enough, but what made the matter worse was the Camera's tendency to behave badly when some scientist or scientists representing an important institution were investigating de la Warr's claims. The reason for this became apparent later.

CHAPTER ELEVEN

WHAT MAKES THE CAMERA WORK?

> Frequently consider the connection of all things in the universe and their relation to one another. For in a manner all things are implicated with one another, and all in this way are friendly to one another; for one thing comes in order after another, and this is by virtue of the active movement and mutual conspiration and the unity of the substance.
>
> MARCUS AURELIUS ANTONINUS

THE fact that if a patient turns his body rapidly while a photograph is being taken of him with the Mark I Camera the picture is blurred, is not due to the same cause which blurs an action photograph taken with a time exposure by an ordinary camera. The patient can make any movement he likes provided it is not a rapid rotational one and it will not affect the image. The blurring, in fact, is not connected with linear change of position but with resonance. This was made clear in another series of experiments.

Since there is resonance between a patient and his blood specimen, it was reasonable to suppose that there is resonance between a plant and its extruded juice, or between a plant and its torn-off leaf. And because magnetic fields are involved, it might also be that resonance is affected by the way in which a leaf is rotated or oriented after it has been removed.

To test this, an experiment was carried out in which a leaf from a small plant was mounted on a card and then placed on a dark slide in contact with an infra-red film. The slide was then placed on the ground and allowed to remain in this position in a closed position so as to exclude the light.

The leaf and slide were rotated through 360 degrees, and at each interval of fifteen degrees an exposure was made, using a new film, making twenty-four exposures in all.

It was found that there was only one position in which the film had been fogged, and this was evidently about the position in which there was resonance between leaf and plant.

To make doubly sure, the leaf and slide were again placed in this critical position. The plant was then dug up and rotated thirty degrees. When another photograph was taken there was no fogging of the film.

There seemed to be no doubt, then, that there was resonance between plant and leaf when they were in the correct rotational positions, though they were separated by a distance of at least 100 yards. Subsequently it was found that this distance could be increased to 200 miles without affecting the result.

When juice from the plant was smeared on a piece of filter paper it was found that the filter paper could be oriented in exactly the same way as the leaf to obtain resonance with the parent plant.

Experiments also showed that a plant was in resonance with the Earth's magnetic field when it was in the position in which it had grown from a seed, for if it were dug up and rotated at intervals through 360 degrees, exposures gave the strongest reaction at the growth position.

To confirm this, young plants were dug up and replanted, and those which were replanted in the original growth position grew at the outset more strongly and rapidly than the others. This is an interesting explanation of what is known as 'green fingers'. People with this gift instinctively replant in rotational positions which favour optimum growth.

The same thing evidently holds good with respect to a patient and his blood specimen, although it is not such a potent factor as in the case of a plant and its juice, because human blood radiates much more strongly. For the best results, the blood soaked into a piece of blotting paper has to be correctly oriented, and there is a rotational position for the patient himself which gives the strongest reaction with his blood spot.

It may seem a strange notion that even liquids such as blood and extruded juice can have critical rotational positions, but it appears that this depends upon the aggregate effect of the countless numbers of the molecules in the liquid, each of which has its directional beam.

Some of the experiments in photography, such as diagnosing a distant patient's physical condition and obtaining a clear

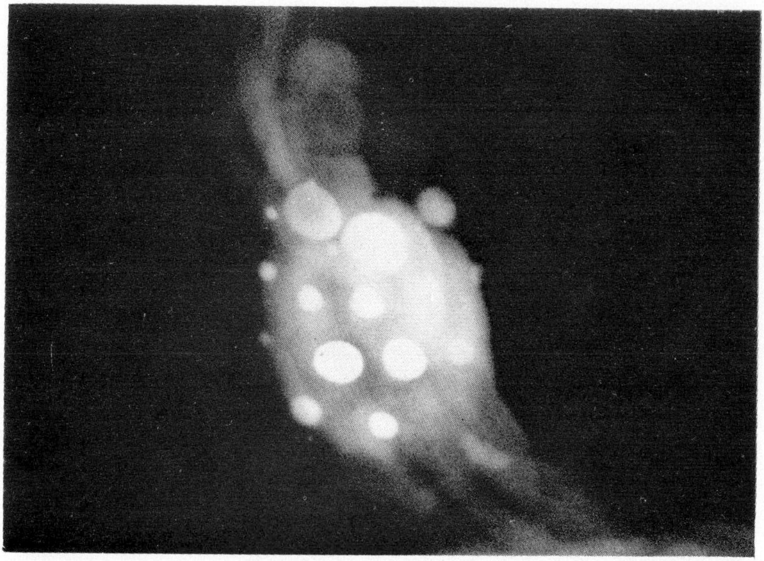

FIG. 24. Radiations from lily seed showing 'bulb' part of life cycle

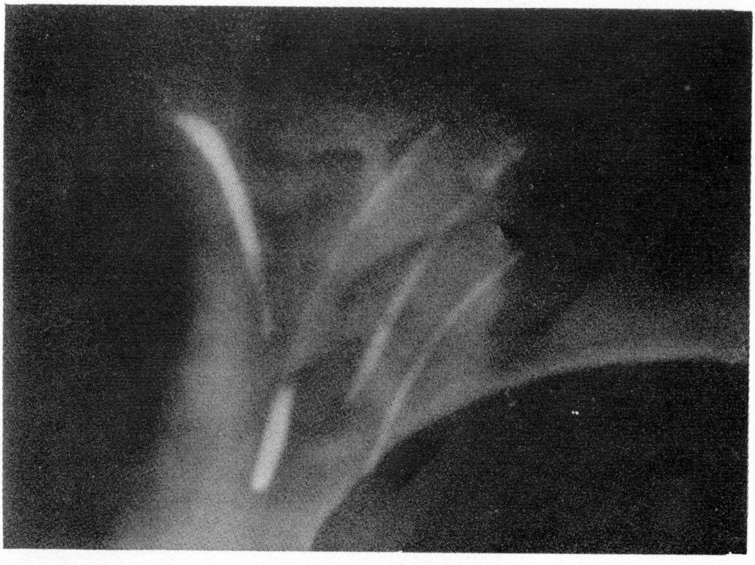

FIG. 25. Radiations from lily seed showing 'flower' part of life cycle

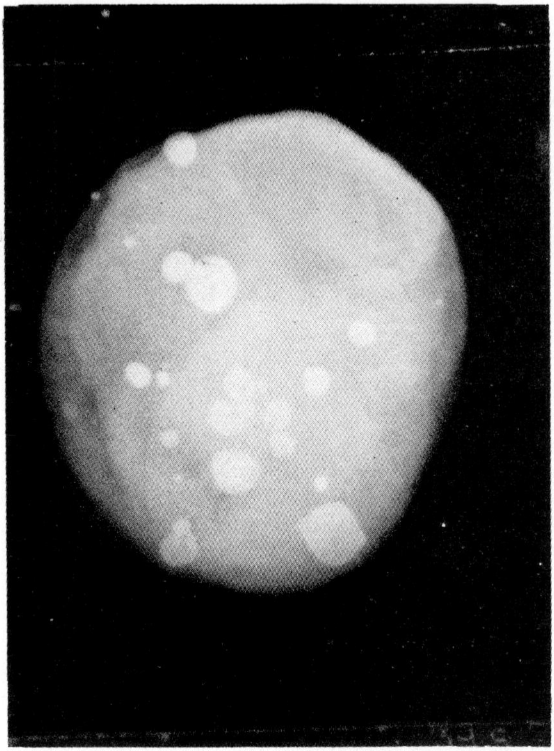

FIG. 26.
Radiations from horse chestnut showing fertility spots

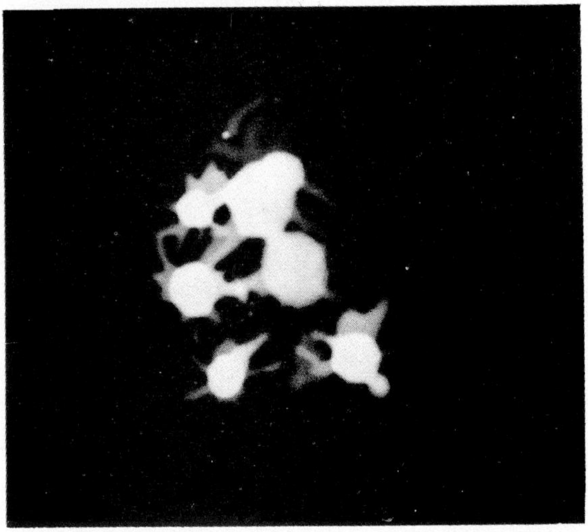

FIG. 27.
Radiations from horse chestnut showing fertility spots at the flowering stage of the life cycle

FIGS. 28 to 31. Four photographs taken on the same day of the same pregnancy case

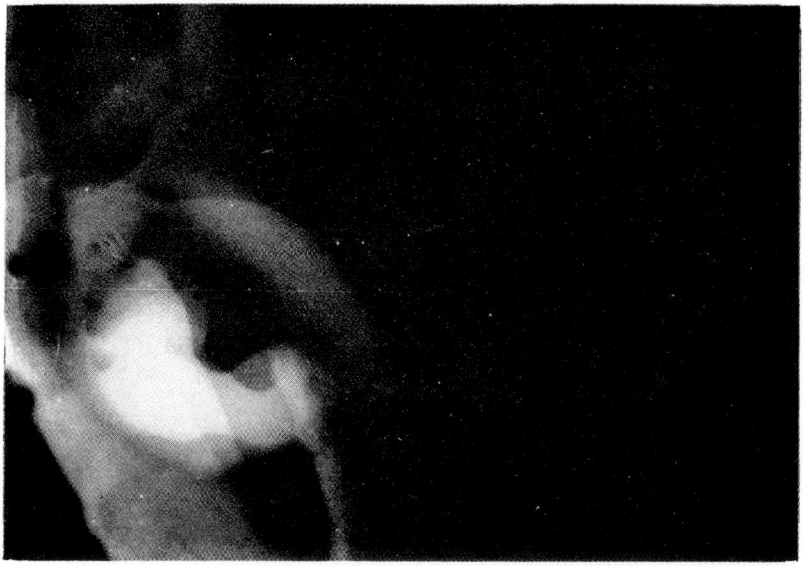

FIG. 28. With the Camera tuned to detect a 1 month pregnancy

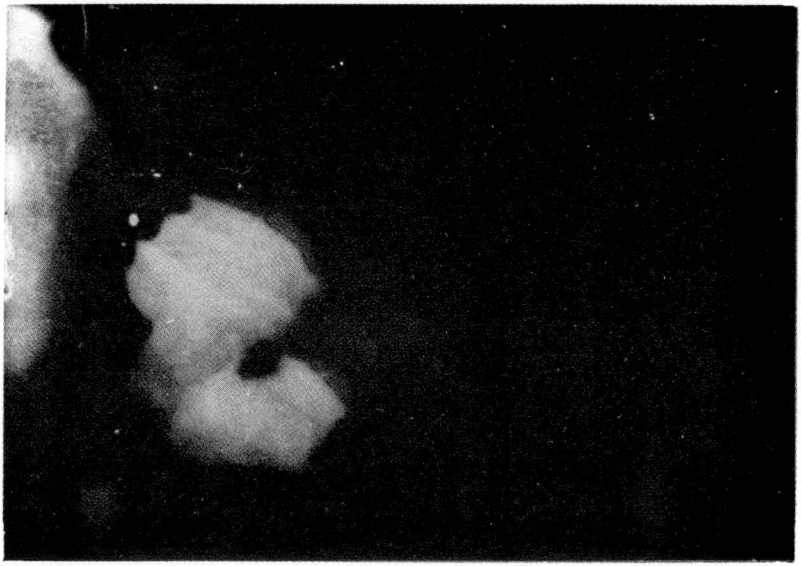

FIG. 29. With the Camera tuned to detect a 2 months' pregnancy

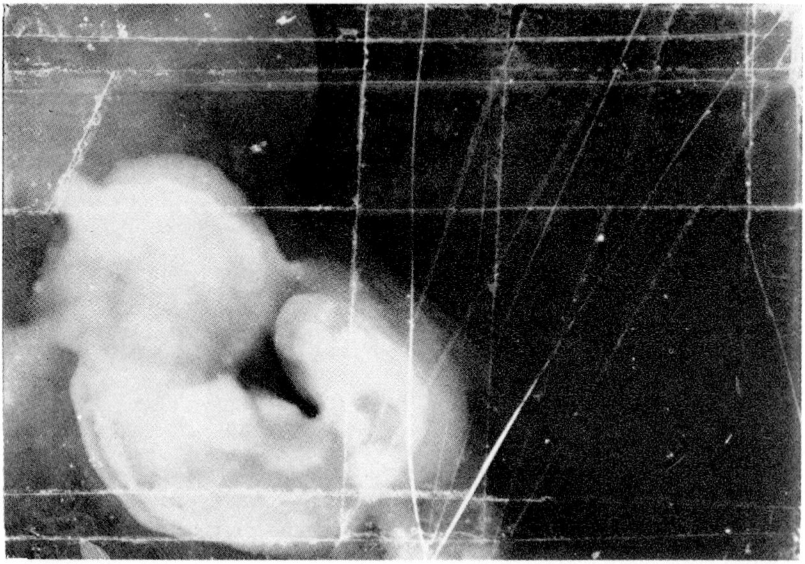

FIG. 30. With the Camera tuned to detect a 3 months' pregnancy

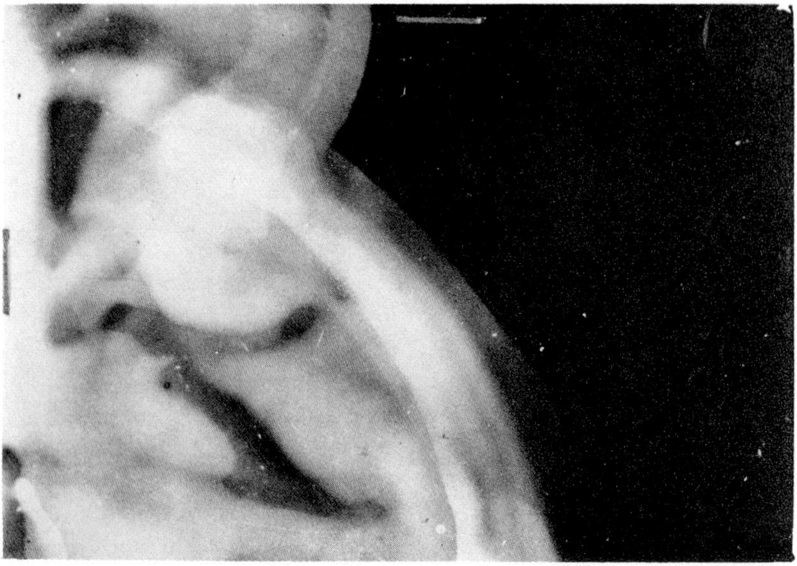

FIG. 31. With the Camera tuned to detect a 6 months' pregnancy

picture of an embryo over an intervening distance of perhaps 100 miles, showed that the Camera is not limited by space in the same way that, for example, electronic instruments are. In one experiment a patient in New Zealand was 'photographed' quite well from his blood specimen which was in Oxford. All physical types of energy such as those on which electronic instruments depend fall off in their intensity as the square of the distance from the source, but the Camera depends upon other kinds of energies which come under different laws.

In the same way it seemed possible that the Camera was not bound by Time as an ordinary camera would be, in which case it might be possible to photograph events in the past. And this would fit in quite well with the conception of the magnetic spectrum which appeared to be multi-dimensional. In addition to the three spatial dimensions of length, breadth and height, the spectrum should have a fourth, or time dimension, and the time vector should be expressible as a Fundamental Ray tuning.

In 1950 de la Warr decided to put this to the test by attempting to take a photograph of his own wedding which had occurred in 1929. He made an appropriate apparatus with an additional tapping on the spiral for the position of the time vector, and with his wife's and his own blood specimens side by side he then rotated the dials, holding the clear thought in his mind: My Wedding Day, 1929.

The portable detector picked up a reaction very clearly. It was so sudden that he was inclined to mistrust it and to search elsewhere with the dials. But after several attempts to disprove the tuning he was satisfied that it was the correct one, and so he exposed the plate.

When developed, the plate showed two dark energy-patterns with a very rough resemblance to human beings. This was such an important result, with consequences so far-reaching, that he decided to repeat the experiment with 'controls'. Two control photographs were to be taken off-tune, both of which should give a nil result if the experiment were valid.

Beginning again, he took a fresh Fundamental Ray tuning for 'My Wedding Day, 1929' and found it practically identical with the setting in the first experiment. He then exposed the plate, took the two control photographs and developed all three of them. It was a gratifying moment when the plates emerged from the fixing bath, one with a clear outline of the same energy-patterns and the other two blank.

Examining the first, he saw that there were two force fields belonging, so it seemed, to two standing figures. Who were they? Using the photographic plate in the same manner as a blood specimen (for a link) and testing with the Mark I Camera, he checked up on the fact that these two force fields were those of his wife and himself.

They were rather crude in outline, but so was the apparatus, which was a mere 'lash-up' of spare parts. Nevertheless, the fact remained that, however imperfectly, *a past event had been photographed*, proving that time was a vector of the magnetic spectrum and that the spectrum had a place in itself for events.

Another very interesting experiment in the new photography threw further light on the principles upon which the working of the Camera depends.

This arose from the somewhat startling photograph which had been obtained from the reliquary of St. Ignatius. A clerical friend offered to perform a ceremony of consecration, and it was decided to photograph a bottle of holy water.

Some Oxford tap water was thereupon ceremonially blessed and photographed, with the result shown in Fig. 32.

Comparing this with Fig. 21 which is the same tap water before consecration, it is easy to see that vital energy has been added to it by the ceremony.

This success prompted de la Warr to carry out a more elaborate experiment using salt in the ritual. This time another priest, robed in cassock and stole, performed a fifteen-minute Service of Consecration according to the Sarum Missal, with the striking result shown in Fig. 33.

Here was something which outshone all previous experiments and which showed clearly the existence of forces and influences not to be found in the scientific text-books. What is the explanation of it?

Subsequently it was found that the pattern on the plate accorded with the thought of the priest while he was performing the ceremony. He admitted that he could never visualise God as a person, but preferred to use a symbolic thought as a stepping-stone to the conception of Divine Power. This had taken the form of a crucifix slightly in perspective, with the head of Christ to one side in the traditional posture and with thongs at His feet.

This may suggest to some readers that the Camera merely photographs what is in the operator's mind. It is not so. Although thought is a contributory factor in diagnosing a disease with the diagnostic instrument, it is by no means the creator of it. It can be categorically stated that in the course of taking over 10,000 exposures no indication has been found that the Camera photographs only what the *operator* knows or is deliberately thinking.

At the same time de la Warr's work seems to show that 'thought forms' do exist. They are real things with no positive mass. Their mass is negative.

Also the physical presence of the operator appears to be essential. To test this, de la Warr fitted a remote-control switch to the Camera and attempted to take a photograph with the operator twenty feet away from it. The result was a blank plate.

Another experiment was devised to find out how much the operator had to know about what the Camera was tuned to detect. While at London Airport awaiting his daughter's arrival from America, de la Warr telephoned the doctor at his Laboratories in Oxford, asking for the Camera to be set to detect his Degenerative Osteo-Arthritis, but without the operator being told what he was to photograph.

An exposure was taken with the doctor absent from the Camera-room, and the result was a very poor photo.

The doctor then telephoned the Camera-room and told the operator what the tuning was. Another photograph was taken, and this time the image was extremely good.

How are we to reconcile this with what has been said before? A tentative explanation is that the Camera is an artificial aid to what is known as Extra-sensory Perception, or E.S.P.

E.S.P. has been exhaustively investigated by Dr Rhine, of Duke University, U.S.A., who has performed hundreds of thousands of experiments reducing the results to statistical analyses.

In these experiments large numbers of people were asked to

name cards which were turned up one by one in another room, describe objects handled by assistants hundreds of miles away, and in general to give information of which they could not possibly be aware. And the results showed that on the average they were able to do this with an accuracy far above that which can be attributed to mere guess-work. The inference is that we possess E.S.P., a faculty poorly developed no doubt, but none the less real, which is able in some measure to operate beyond the limits of time and space.

It seems that in using the Camera the E.S.P. of the operator probes the force field of the patient—or whatever the subject of the photograph may be—and the Camera records the resulting change in the *operator's* force field.

At the same time the operator's ordinary mind plays its part. In Dr Rhine's tests it was necessary for his subjects to know what they were being asked to reveal, and in the same way the operator of a Camera must have in mind the thing he is seeking to photograph. The ordinary mind acts as a sort of link; it does not do the probing.

At this stage is it impossible to be dogmatic. But as we shall see later, the results which can be obtained by this new and strange form of photography are too closely related to verifiable results to be dismissed as imaginary pictures in the operator's mind. The photographs always correspond to something which exists, even if it is only thought forms created by an officiating priest, but the mind of the operator is a necessary factor in creating the conditions which enable them to be taken.

CHAPTER TWELVE

TREATING PLANTS THROUGH THEIR PHOTOGRAPHS

> It is in a dark and cold world we sit, if we will not open the inward eyes of the spirit to the inward flame of nature.
>
> GUSTAVE TH. FECHNER, *The Soul Life of Plants*

DE LA WARR had discovered in 1950 that the emulsion on an ordinary photographic plate was linked in some way to the person whose photograph was taken. In the same way, it was found that a plant was linked to its photograph. This being so, it was to be expected that the growth of a plant could be stimulated by irradiating its photograph with a radionic device such as the Colorscope. Further research work showed that this appeared to be the case, and in 1954 a series of controlled experiments on plants carried out at the Delawarr Laboratories proved it beyond doubt.

How can there be a link between a plant and its photograph? If we photograph a cabbage, the image on the plate is formed by the multiplicity of reflected light-rays coming from the cabbage. But besides these light-rays, vital radiations of the kind which have been discussed in this book are also emanating from the cabbage and these, too, are received by the emulsion.

At risk of repetition, each molecule of matter can carry an electrical charge which is *specific* for that particular molecule when it is charged in this way; and this charge can act as an inconceivably minute radio station which both transmits and receives its own particular signals. When all these countless myriads of charged molecules are broadcasting, they build up what may be called a 'generic pattern', which, as we have already noticed, is the means whereby form or shape appears in the material world.

The combined signal from a plant or a human being, a signal

which is built up by all the separate broadcasts of the billions of charged molecules which compose it, is unique. This cabbage in your larder is not quite the same as any other cabbage; Mr Brown is a little different in his physical make-up from any other man in the world. In the same way, the signals broadcast from this cabbage, or from Mr Brown, are different from any other broadcast signals.

Furthermore, since each tiny molecular charge is also a receiving station, the generic pattern of a plant or a human being governs the pattern of signals received from outside. This is where a photograph plays its part. The emulsion retains the generic pattern of the thing which is photographed, and therefore it acts as a sort of tuned transmitter. If a radionic broadcast is projected through it, this generic pattern will transmit exactly the pattern of radiations suitable for affecting the plant at a distance.

Because plants are closely linked to the soil in which they grow, and especially to the micro-bacteria which form a vital part of the soil, it seemed probable that the soil itself contained prodigious numbers of molecules carrying electrical charges and that these charges affected the soil's fertility; also that the rise and fall of the micro-bacterial population was in some way related to the increase and decrease of the electrical charges. It was also likely that the charged molecules in the soil were not evenly distributed, but lay in pockets.

A number of experiments were carried out, and it was found that two copper test prods connected to a galvanometer gave alternately high and low readings when pushed into the ground. The readings varied a good deal with the site, and whereas they were highest when the prods were first inserted, they gradually decreased and fell to zero after a few hours.

There was a great difference in the electrical conditions in undisturbed soil and in soil which had just been dug. When the soil was removed from a test hole of about three feet square and nine inches deep, mixed thoroughly and replaced, the galvanometer readings were the same at different depths, but they were extremely low. This showed that digging and mixing had the effect of discharging the pockets of electrical potentials and distributing the charges uniformly—a fact which should interest the school of thought which believes that better crops are to be had by leaving the soil undisturbed.

A more detailed series of experiments was carried out to measure the electrical charges in soil which is left in its natural state. The apparatus used was a brass pot, in duplicate, to which was fixed a central brass electrode on an insulator, the insulator preventing any contact between the electrode and the pot.

The pot was suspended from the galvanometer to avoid any leakage, and the earth connection of the galvanometer was connected to the central electrode. The pot was then filled with soil which was kept moist during the experiment.

From July 29th, 1954, until September 22nd readings were taken every two hours. They showed very considerable variations. Once they ranged from seventy-seven microvolts minus to twenty-two microvolts plus in twenty-four hours.

What was the cause of these variations? Recordings of rainfall, temperature and sunshine which were carefully made showed no signs of any relationship with the electrical readings; nor did it seem possible that electrolytic action between the soil and the metal surfaces could explain the phenomenon. Local electrolysis in the soil seemed nearer to the true explanation, although it could not fully account for the results. The most likely answer to the problem was that the myriads of local micro-bacteria are subject to rapid changes and that these changes affect the electrical charges in the soil.

If this were so, then radionic treatment which can influence living cells can also influence the micro-bacteria, which in their turn will produce the electrical stimulation to promote greater fertility. On this hypothesis another experiment was carried out.

Two sites forty feet apart were chosen close to the Laboratories. To eliminate any possible variations in the soil, the soil was removed from each site, sieved, mixed together and replaced. After six days in which the soil was allowed to settle, Site A was photographed with a standard plate camera. Site B, the control site, was not photographed.

For one month Site A was treated radionically by irradiating the photograph daily in the Laboratory dark-room. Each site was then planted with four cabbages which as far as possible were precisely of the same size and sturdiness.

For the first two weeks there was no apparent difference in the growth of the two groups of plants, but from then onwards until the end of the experiment the cabbages in Site A (Fig. 35) grew at least three times as fast as those in Site B (Fig. 36). Four weeks from the time of maturity, the treated plants were in fact about three times the size of the untreated controls.

Arrangements were made to repeat the experiment on a larger scale. This time a continuous strip of garden, seven feet in width and thirty-seven and a half feet long, was chosen because three rows of peas already growing in it were so uniform in size that they proved an unusual consistency of the soil.

Sixteen plots were laid out. Fifteen of them, including the untreated buffer plots in between the experimental ones, were exactly the same size, while Plot 16 which ran right across the site acted as an additional control. The fifteen plots included six treated plots, two untreated control plots and seven buffer plots.

The random selection of the plots was carried out under the scrutiny of Dr. E. W. Russell of the Soil Science Laboratory, Oxford University.

Two of the plots were photographed and the photographs were irradiated in the De la Warr Laboratory every day for a month. Ninety-one broccoli plants, about seven inches in height, were then planted eighteen inches apart over the whole of the garden strip. Four more plots were photographed and added to the treatment list.

The experiment came to an end when severe frost and snow had apparently stopped all growth. After sorting and weighing, it was found that the treated plants were eighty-one per cent heavier than the untreated control plants. The photograph in Fig. 37 which was taken in October shows the relative sizes of the treated plants (on the left of the picture) and the untreated control plants (on the right of the picture).

No necrosis could be discovered on the treated sites, but some was visible on the control sites.

Previous work in the Laboratory had shown that a plant in its original growth position is also in its critical rotational position and therefore in resonance with the life sources which sustain it: in this position it is receiving the optimum quantity

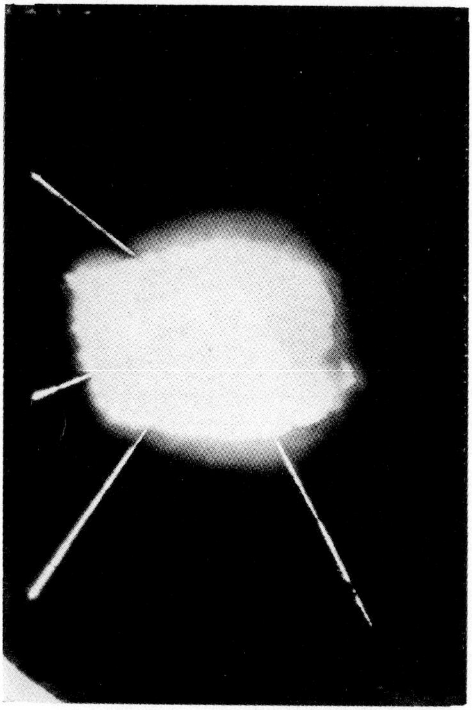

FIG. 32.
Radiations from
Oxford tap water after
ceremonial blessing
by the
Rev. J. C. Stephenson

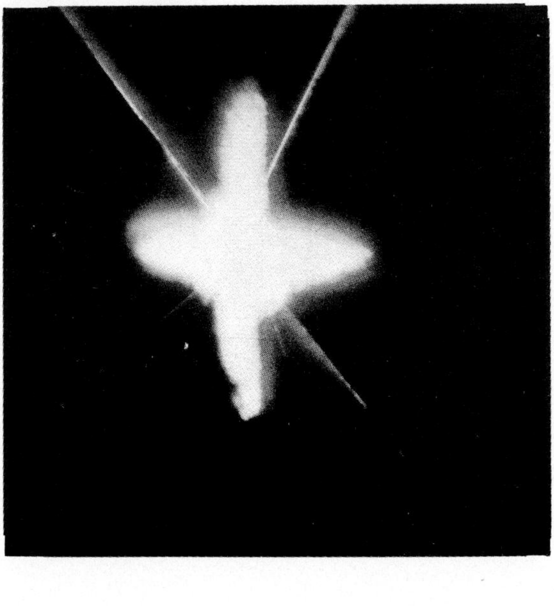

FIG. 33.
Radiations from
Oxford tap water after
ceremonial blessing
by the
Rev. P. W. Eardley

FIG. 34. Brass pots suspended from galvanometer

Fig. 35. Treated Cabbages

Fig. 36. Untreated Cabbages

FIG. 37. Treated and Untreated Plants

of vital radiations. In the same way, the soil contained in a flower-pot will gradually acquire a similar characteristic so that if the pot is rotated, there is a critical position which is the best for growth. When in this position it is also in resonance with other pots of soil of the same size which are in their critical rotational positions.

To demonstrate this, three pots were used in an experiment arranged between London and Oxford. After being filled with London soil, Pot 1 was rotated to its critical position with the aid of the Portable Detector, while Pot 2 was rotated to its critical position in relation to Pot 1. Pot 3, the control pot, was rotated off-tune to forty degrees out of its critical position, and an odd shape of pot used. Pot 1 was then photographed with a Leica camera and the negative sent to Oxford for radiation purposes.

Three broad bean seeds were planted in each pot and given daily treatment by irradiating a photograph of Pot 1 only. After 16 days, germination was apparent in all three pots, but in Pot 3 only one plant was visible, whereas in the other two pots all three plants had appeared.

At the end of the experiment the heights of the plants in Pot 1 were eight inches, eight inches, and four and a half inches; those in Pot 2 were eight inches, six inches, and four and a half inches; while the single plant in the control pot was only five inches.

In 1955 a series of experiments on farm scale were initiated, in districts as widely separated as Fife (Scotland), Yorkshire, Shropshire, Cheshire, Gloucestershire and County Dublin. Some of these experiments are being conducted under strict control conditions, and the de la Warr instruments, including the Colorscope, are being used for the purpose.

Although some encouraging results were obtained it would be premature to advance any specific claims until, in due course, an official report can be issued which will embrace the whole of this important series of experiments, which are still in progress under a very wide diversity of conditions of climate, soil and rainfall.

Five acres of a tobacco farm in Rhodesia were treated by a rather different method. In this case no photos were used but pinches of top soil were collected at 10-foot intervals and sent to Oxford for Colorscope treatment.

In his anxiety to get good results and being unable to watch the effects, de la Warr over-treated with the result that the tobacco plants flowered before those on any of the neighbouring farms. This was bad because tobacco farmers want leaves not flowers. However, the quality of the leaves of the treated plants proved better than that of the untreated, although they were too small, largely due to a 3-weeks' drought at the time. The crop got only 5 inches of rain against 8 inches elsewhere; but the farmer was impressed, for treatment had an obvious effect, and he wishes to try again next season.

These experiments and many others on similar lines by the Homeotronic Memorial Foundation of Newport, Pennsylvania seem to prove beyond doubt that there is a link between living entities and their photographs. At the present stage there is not enough data for more than a general statement of this extraordinary fact, but ten years of basic and applied research work in Pennsylvania, California, Arizona and Florida have proved that soil, crops and trees can be treated successfully by irradiation through photographs. This independent corroboration of de la Warr's work is important.

The so-called radiotonic process used in America—a phrase coined by the late Dr William J. Hale, former Chief of the Dow Chemical Research—conditions the soil through broadcast radiations. Its primary purpose is to treat living cells through their own amplified generic wave-patterns and also by the amplified wave-patterns of chemical reagents which alter the structure of the cells without damaging them.

By using aerial photographs, some 30,000 acres of diseased or infected trees have been successfully treated. Attempts are to be made now in Florida to eradicate what is called 'the world's worst pest'. This is the nematode, which is sometimes found as much as fourteen feet deep in the soil and which seems to be impervious to insecticides. In Florida many citrus and pine trees are said to be doomed unless some means can be found to deal with the pest.

PART II

CHAPTER ONE

THE PITFALLS OF A PIONEER

> Science has organised this nervousness into a regular technique, her so-called method of verification; and she has fallen so deeply in love with the method that one may even say she has ceased to care for truth by itself at all. It is only truth as technically verified that interests her. The truth of truths might come in merely affirmative form, and she would decline to touch it.
>
> W. JAMES in *The Will to Believe*

SCIENTIFIC discoveries can be divided roughly into two categories. Those which arise from an extension or development of knowledge which is accredited and to be found in the text-books, and those which, following an original line of their own, appear to contradict the accepted tenets.

Discoveries in the first category are far less apt to meet with incredulity or opposition since they are in agreement with current trends of thought. They fall on prepared ground, and the organised forces of Science are usually willing to give them support and acclamation.

But those which are contrary to established scientific concepts are in the position of heretical creeds in the face of religious orthodoxy, and the men who make them often have a sterner battle to win acceptance of their work than they had in their laboratories.

It should not cause surprise, then, that de la Warr has not trodden a primrose path in his contacts with many of the doctors and physicists who have investigated his claims. In describing what has happened during the last ten years there is no intention to criticise these men. Heaven knows, they must have found it hard even to consider the possibility that such apparently fantastic claims might be true.

But if this book is to be the 'documentary' which it sets out to be, it is absolutely necessary to include some account of how these ideas have been received up to date by the outside world.

And all the more so because the impact of critics and investigators has not only given de la Warr's work an extremely severe testing, but it has instigated further experiments and revealed strange and most unexpected factors which have added to the knowledge already gained.

In 1944 before the Laboratory had been built and when research work was carried out in the bath-room, Sir Farquhar Buzzard was persuaded to come down to Oxford and see the diagnostic instrument. The lounge was fitted up as a show-room.

Sir Farquhar arrived with two colleagues, one of whom was a doctor working in the Oxford Pathological Laboratories. He listened attentively to what de la Warr had to say and watched him give a demonstration of testing a proposed remedy for some disease, namely penicillin. Then he asked de la Warr what he would like him to do and he offered to bring along some patients on whom he could carry out some tests.

All this time the doctor from the Pathological Laboratories had been fidgeting, and finally he rose and stared out of the window as if the proceedings no longer interested him. When Sir Farquhar asked him what he thought of the demonstration he turned to de la Warr and said: 'I can't understand what you are doing and I can't accept any of it. The very way you handle the penicillin shows that you know nothing about it.'

This was direct enough, but Sir Farquhar turned it by asking him what tests he would prefer. The doctor replied: 'We will send twelve bottles labelled only with numbers, and if de la Warr can tell us what is in them we will send twelve more. It may take him ten years to prove his case.'

De la Warr explained that inert samples of inorganic matter in bottles lacked the vitality which was present in blood specimens, and so on, and that the diagnostic instrument was only designed to probe the force field of a living patient. This explanation was regarded as an excuse to evade a test and the matter was dropped.

These early days were difficult for de la Warr. In 1945 he lectured to the London Medical Society for the Study of Radiesthesia, a scientific body whose members are qualified medical practitioners, but although his lecture was well received, it did not set the Thames on fire.

Perhaps if he had allowed people to look inside his diagnostic instrument he might have won over more practitioners at this stage, but on the other hand he might have shattered the credulity of those who expected to find complicated electronic apparatus. This question of seeing inside his instruments was always a difficult fence to take, because apart from the bar-magnet the various components were meaningless to orthodox science. Any orthodox scientist or engineer who examined the simple make-up would probably have considered the whole thing a fake.

Some little time later a chance came of demonstrating the effectiveness of his acoustic treatment set. In 1946 the medical profession was interested in a comparatively crude method of stimulating the pituitary gland by short-wave therapy, a method introduced from Holland by a Dr Samuels. His was an expensive apparatus, and although it appeared to have beneficial effects on certain types of disease, it was highly dangerous in de la Warr's opinion when it was used to irradiate the brain.

A doctor who knew of de la Warr's work wished to discover if his treatment set could produce the same effect as the Samuels apparatus, and so a comparative test was made at this doctor's address in London.

The conditions of the test were to irradiate a patient's pituitary gland for a certain length of time and then examine the spectrum of his blood through a spectroscope and see whether some specified change in it had occurred. De la Warr irradiated the patient for ten minutes with his set, which incidentally cost a mere fraction of the sum charged for the Samuels apparatus, and on examining his blood it was found that this particular change *had* occurred. The doctor was delighted with the result, but so far as de la Warr knows, he was unable to arouse any interest among his colleagues.

It would be unfair, perhaps, to criticise the medical profession on this score. Dr Samuels was a qualified medical man, and short-wave therapy comes within the province of classical physics, whereas the radiations investigated by de la Warr fall outside it.

The doctors were certainly not slow to realise the possibilities of X-ray treatment when Roentgen introduced it in the 1890s. And this in spite of the secret having leaked out before its official announcement and the most absurd stories and speculations having been circulated. An English merchant had been advertising X-ray proof garments for modest ladies, and a Professor in New York had talked about using X-rays to penetrate the thick skulls of slow-witted students and project knowledge directly into their brains.

But X-rays, of course, figure in the text-books, whereas the energy with which de la Warr is concerned has sometimes been alleged to be too 'subtle' for laboratory demonstration. But this is not true. Anyone who possesses the necessary knowledge and apparatus can demonstrate these forms of radiation in the laboratory.

However, as has already been explained, these wave-forms are in a different category from the energies of the Electromagnetic Spectrum, falling into the same class as the 'radiations' from a magnet.

Was it possible to demonstrate some firm connection between these two classes of energy? If only this could be done it would make these wave-forms 'respectable' in the eyes of orthodoxy by giving them a place in the scientific genealogical tree. The scientist, of course, would like to confine any tests he makes to the use of his existing range of apparatus whereas it is obvious that the apparatus for detecting these subtle radiations has not yet been perfected. The de la Warr Camera is the first piece of apparatus for the purpose yet discovered that will produce repeatable results in this new field.

Meanwhile favourable reports continued to come in from practitioners both qualified and unqualified who were using the Colorscope, and the idea occurred of designing pre-set instruments for various maladies so that they could be used by patients at will.

A pre-set treatment instrument was devised for fibrositis, incorporating the use of an infra-red lamp instead of a simple light-bulb. The mechanism in front of the lamp was tuned to treat fibrositic tissue, and also bone marrow since this last is often the source of arthritic or rheumatic conditions. The results to date have been most satisfactory with very few exceptions, but it is found that certain patients need *specific* tuning to meet their individual cases.

FIGS. 38 and 39. Results of Broad Bean Experiment

With the small number of Colorscopes and other apparatus at present in use, about 300,000 treatments are being given each year. This is a mere nothing in comparison with what could be done, and in 1951 the de la Warrs decided to set about opening a regular practice in Oxford. Their treatment centre was to be run by a fully qualified doctor engaged by the Laboratories which by now was a limited liability company. This doctor would have the service of a number of operators trained to diagnose by the de la Warr methods and to give treatment by Colorscope, or by continuous broadcasts, according to the circumstances.

Alternatively, patients would be allowed to hire a smaller type of Colorscope and use it in their own homes under medical supervision. The clinic would be equipped with a Mark I Camera for confirmatory diagnosis and for progress photographs to check the development of each case. A service would also be run to help medical practitioners in this aspect of their work.

It was a good idea, and if put into effect it would be the first clinic in the world which claimed to be able to diagnose photographically and treat specifically by these methods. It would give new hope to patients suffering from tuberculosis, cancer and other grave maladies which do not respond readily to ordinary methods of treatment. Of course it would mean a very great deal of work for the de la Warrs and their staff; but it would be rewarding work and it might give a fresh stimulus to their research work which by now was languishing for lack of funds.

An elderly doctor agreed to consider accepting the post of medical superintendent, terms were discussed, and a suitable three-storied house was found. Only a lease-hold purchase was possible since the property was owned by St. John's College.

The next problem was to raise the money needed for a thirty-five year leasehold tenancy plus the cost of equipping and running it for the first year. A mortgage was applied for and the first difficulty arose when the building society refused to consider a mortgage on leasehold property. After some consideration it was decided to include the de la Warr's own home in the application, and the building society eventually agreed to advance a certain sum repayable within twenty years.

The de la Warrs were now heavily involved, but they had faith and a clear purpose. Their agent made arrangements to close the deal. But at the last moment he reminded them that they must get permission from St. John's College to use the premises as a treatment centre.

When de la Warr explained the situation, the Bursar grew nervous about the probable reaction of the other doctors in the neighbourhood to the centre; he was unwilling to agree to the proposal until he had consulted them. After some delay he told de la Warr that he was not prepared to put forward his application unless the doctor in charge of the clinic was approved by the local doctors.

De la Warr wrote to the doctor in question and explained what had happened. He was dismayed when this doctor replied saying that he had reconsidered the proposition, that he felt the British Medical Association might take exception to his working with unqualified persons even as a limited liability company, and that he could not therefore go forward with the plan. This brought the whole project to an end.

Other difficulties beset the de la Warrs', this time it was in obtaining patent rights in America for his Colorscope and his Camera. The United States Examiner was particularly puzzled that he should claim any value in the critical orientation of a bar-magnet or the use of a spiral with adjustable tappings, since no previous work had been done in this field. Affidavits were sworn by practitioners who had used these instruments successfully, but after protracted and expensive argument the patent rights were refused.

CHAPTER TWO

HOPE SPRINGS ETERNAL

> The major part of the scientific world does not *wish* to examine the evidence, but endeavours only to evade and escape from it. It is not animated by a scientific desire to know the truth but is in the grip of a psychological urge to disallow what is distressingly unfamiliar.
>
> G. N. M. TYRELL in *The Personality of Man*

FROM the Laboratory on the hills overlooking Oxford there is an exhilarating view of the city's spires. Here was one of England's great seats of learning. Surely it should be interested in the remarkable discoveries which were being made on its very outskirts?

De la Warr had taken a number of striking photographs with his Camera, photographs which showed beyond doubt that he was on the track of new knowledge, unknown laws. All over the Kingdom there were men of brains and ability who could easily verify the results of his work and help to develop this new branch of science. But how could he attract their attention and get a hearing?

He wrote to the Royal Society asking if it would accept a paper, or some communication, about his work. The reply was that unless such a paper was sponsored by a Fellow of the Society, nothing could be done.

So he telephoned the President of the Society, briefly described his work and asked him if he was interested. The President replied that he himself was a chemist so that the work was a little out of his line, but he suggested contacting Professor L. who would probably take the matter up.

De la Warr did this, and in due course Professor L. came to the Laboratories and was shown everything except the inside of the Camera. He was extremely interested but not very hopeful. The immensity of the discoveries bewildered him and he doubted if such things could be 'got over' to the average physicist.

He suggested that de la Warr should make a very simple piece of apparatus which would merely photograph the Fundamental Rays of the elements and that he should offer to lend it to Oxford University. If this were done he thought someone might be persuaded to investigate it, but he warned him to say nothing about the revolutionary theory of Cosmic Energy which by this time de la Warr had formulated—a theory which will be described later.

This seemed good advice, so de la Warr made the apparatus and again contacted Professor L. who promised to send over a woman crystallographer, a Fellow of the Royal Society, to talk the matter over.

Once again hope rose high. On the appointed day de la Warr called at the University to meet Mrs M. and drove her to his Laboratories along with her small child.

Nothing could have been more disappointing than the upshot. Mrs M., a vague, unreceptive lady, seemed to be incapable of understanding de la Warr's work, and at length when her patience and that of her child who for some time had been tugging fretfully at her skirts, was exhausted, de la Warr drove them back in chilly silence. Her last words were that she did not think the University or the Royal Society would be interested in what he had shown her.

After this set-back de la Warr decided to press on with the medical side of his work and market the new Camera in order to help his finances. Research work costs considerable sums, and it is somewhat hard on a research worker when he is obliged to fight a grim financial battle as he goes along. But there was no other alternative, and the readiest means of making enough money to avoid closing down the Laboratories seemed to be the sale of his Camera.

The first thing to do was to embellish the apparatus and make it a marketable proposition. Using polished bakelite sheets and chromium-plated fittings, he produced his Mark I model, which is shown in Fig. 40. The actual 'Camera' part of the apparatus is in the small boxlike structure supported on the four pillars. This camera head houses the lenses and the photographic plate whilst the main cabinet below houses the magnets and tuning devices required to produce the two beams of radiation already referred to.

Standing five feet high and measuring three feet square, it was a most handsome instrument. Knowing what it would do, he saw every chance of launching it successfully on the market.

The occasion seemed to call for a Press reception, and so he approached his local newspaper, the local agent for the News Pool, the *Sunday Express* and the London branch of *Time-Life*.

On the day of the reception an imposing assortment of photographs was laid out for viewing. Two reporters arrived, followed by an American lady with a chauffeur in the biggest car on the American market.

After cocktails, de la Warr explained briefly about his work. The local reporter strove valiantly to adjust his mental processes, and the News Pool man, a terse Scot, asked just enough pointed questions to throw his mind into a confusion from which he never recovered. The American lady, with the intellectual determination of her race, fought her way through to some understanding of what de la Warr 'claimed to have done', as she called it, and then tried to write it down.

They all went away saying that it would be a very difficult matter to write up. However, the *Sunday Express* reporter came back later to ask some more questions, was impressed, and returned to London to try it out on his editor. That was the end of the Press reception, and the de la Warrs waited hopefully for a little welcome publicity.

They were disappointed. The local newspaper gave them half a column of timid reportage, but nothing appeared in any of the other papers.

There was a ray of hope when the *Sunday Express* man came back a second time to say that whereas he and his colleagues were inclined to think the work genuine, his editor was not. Perhaps this was understandable since the editor had consulted an eminent physicist who pronounced the whole thing as quite impossible. However, the reporter was not to be put off, and he had brought with him a fifteen-point questionnaire. This de la Warr answered to his satisfaction and added some points of his own.

At this juncture the reporter mentioned that he had some shrapnel in his lung and he invited a test. So with twenty minutes to go before his train left, a blood specimen was taken and it was rushed into the Camera-room.

The Camera was tuned to 'Foreign body in the lung', and with a fervent prayer that this rate would work, a photograph was quickly taken. Ten minutes later they were all looking at a dripping plate which clearly showed the outline of a foreign body, a dark patch on the edge of the lung's arch.

Driving with the reporter to the station, de la Warr's hopes rose again. Surely this piece of personal proof, so obviously unrehearsed, would bring conviction to the editorial mind? But no, nothing appeared in the *Sunday Express* except a bald statement that the Delawarr Laboratories had detected a piece of shrapnel in a man's lung by means of photography.

No doubt it would be unfair to accuse the editor of undue timidity, for it is hard to accept such apparent miracles as these radiation photographs without a personal visit to the Laboratories. De la Warr thought that the scientific world might be more inclined to listen to him if he offered a small fragment of his discoveries, something so modest that it would not conflict too strongly with the accepted tenets.

So he wrote a simple article dealing with the radiations of the elements and sent it to *Nature*. The editor returned it on the plea of insufficient space. He tried a letter to the British Medical Association, enclosing three of his best photographs. It was returned with a chit to say that the B.M.A. was not interested. The same letter and photographs were sent to the American Medical Association, only to elicit a similar reply.

It is a well-known fact that most scientific bodies have to deal with cranks and impostors who pester them with impractical suggestions which cannot stand up to tests. In self defence they have to erect a barrier against these would-be wasters of their time. De la Warr realised this and he also knew that to many doctors and physicists his photography would seem so fantastic as to suggest pure fraud.

How could he get over this? Suppose he allowed some qualified doctor in a key position to *take the Camera away and use it himself*. Surely he could not fail to acknowledge that it worked, and once the testimony of such a man was on record, the medical profession could hardly fail to investigate his claims?

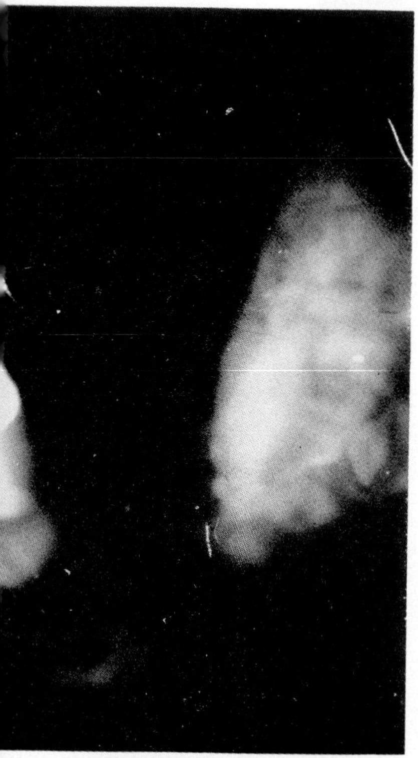

Fig. 41.
Tuberculosis of the Lungs

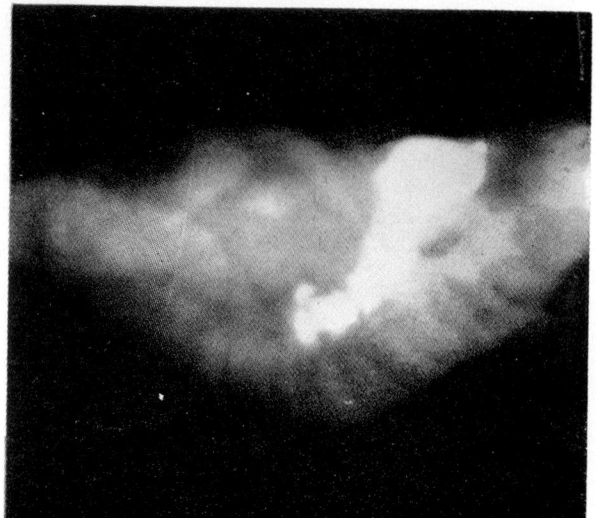

He approached a commercial firm which contacted a young doctor of considerable character and ability on the staff of a big general hospital in London. This doctor consented to meet de la Warr. He examined some of his photographs of disease conditions which had been taken from blood specimens, and overcoming his rather natural prejudices he not only remained polite but even lent a sympathetic ear to the story of how the discoveries had been made—which to any medical man who had never heard of such things must have sounded well-nigh incredible.

Gaining confidence at his first interview, he agreed to visit the laboratories and see the apparatus with his own eyes. De la Warr knew the risk, for his work was largely in the world of atomic physics and most doctors are not very good physicists. But it was a risk which had to be taken.

Some of the doctors he had met were men whose minds were not fully open to completely new ideas, but this one was different. Perhaps the obvious sincerity of the de la Warrs impressed him; possibly the white-coated staff, men who were obviously not cranks, gave him confidence. However it may be, he agreed to take some photographs himself immediately.

Having satisfied himself that the Camera really did work, he began to come down to the Laboratories once a week until he was convinced that the claims made for this new branch of science were just ones; and also that he could operate the Camera himself with the aid of a physicist.

At this point he decided to ask permission from his Chief at the Hospital to import the Camera and keep it there under lock and key. His request was granted.

Some exciting months followed. The de la Warrs made frequent trips to London with their physicist to give what help was needed. Case after case in the hospital was photographed from the patient's blood specimen and the results were checked by X-ray photography, biopsy or autopsy.

More than 400 photographs were taken with excellent results. In a few cases the vibrational rates, which as explained earlier had already been fixed provisionally, had to be modified, but in most instances they were sufficiently accurate, and on the whole things went very well.

Among the first batch of cases chosen from one of the hospital wards was one of tuberculosis of the lungs (Fig. 41), and one of gastric ulcer (Fig. 42).

In Fig. 41 the Camera was tuned to 'Tuberculosis of the Lungs', and the dial setting for this was 402776. In these photographs, which are prints from the plates and therefore 'negatives', the disease radiations from the affected organs appear as large white patches. Notice that this is unlike X-ray photography in which the bones and more solid parts of the anatomy stand out clearly. What is being photographed is distant *radiation*, so that accuracy and clearness in anatomical detail is not to be expected at this early stage of the Camera's development.

Much was learned from these photographs taken in hospital. It became clear, for instance, that three factors entered into the taking and the interpretation of radiation photographs. These are: the vibrational rate which is used on the dial setting, the intensity of the disease radiation, and the anatomical interpretation depending upon the structural features of the radiation pattern.

The third factor is easily the least important; on the other hand, the first factor is vital. As explained earlier, a disease changes its rate of vibration as it develops, and if a series of photographs is taken on slightly different rates, that which gives the clearest picture indicates just what stage the disease has reached.

All the cases chosen for photographic detection in the hospital were subjected to the usual thorough routine tests. The checking up was sometimes done before the Camera was used and sometimes afterwards, so that on some occasions a case was dealt with 'blind'; that is to say, the doctor had no previous certain knowledge of what the Camera should show.

In most cases the results were very satisfactory, but there were some exceptions in the case of bronchial carcinoma. The difficulty here was to establish a vibrational rate which gave consistent results. At length it was realised that there were many stages and ramifications of cancer of the lung and that several rates had to be used before a correct picture could be obtained.

FIG. 40. Th
and Mr St

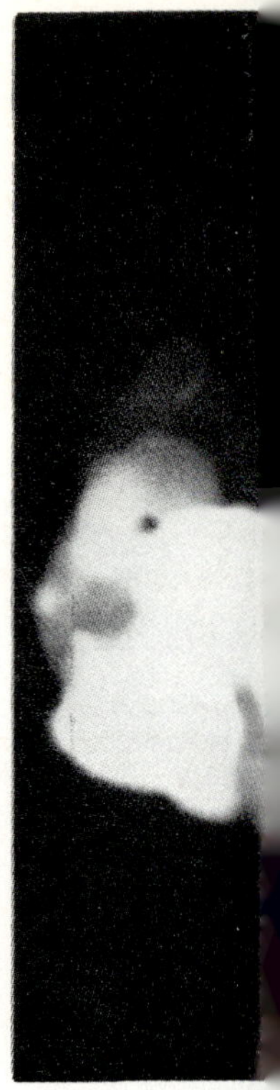

FIG. 42.
Ulceration of th
Stomach

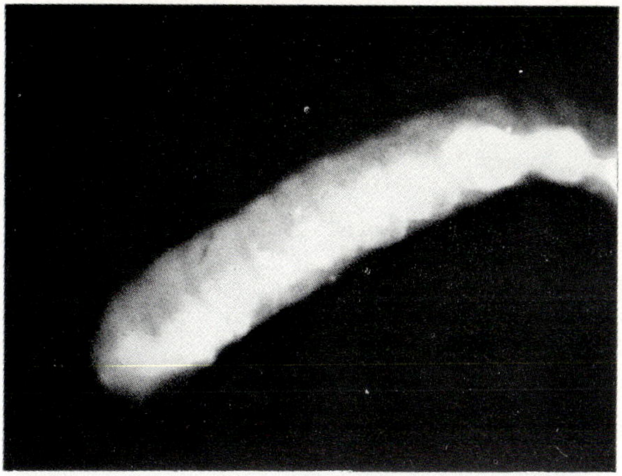

FIG. 43. Buerger's Disease in posterior tibial artery using photograph instead of blood specimen

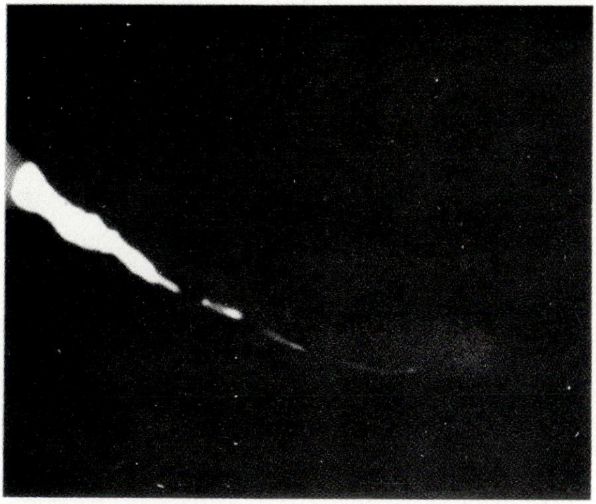

FIG. 44. Buerger's Disease in posterior tibial artery using blood specimen

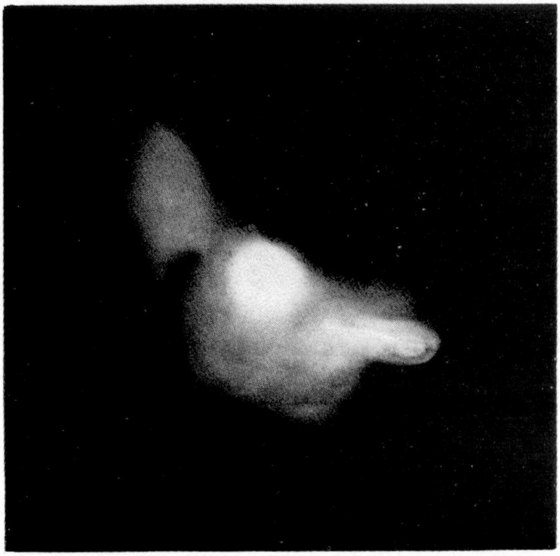

Fig. 45. Iron wire and stone in reticulum of cow's stomach

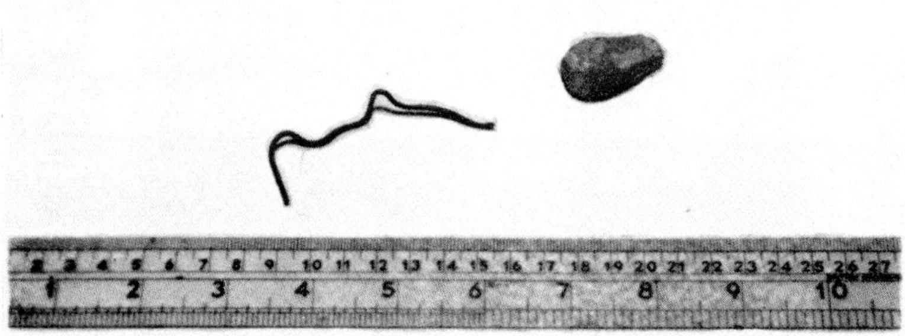

Fig. 46. Iron wire and stone after removal by Mr J. D. Warde, M.R.C.V.S.

The doctor who was using the Camera became more and more enthusiastic as case after case was checked up and found to have been accurately photographed.

When some future historian writes an account of the origin and development of this new technique, he will have to put it on record that the Camera during its first official test was housed in a room hardly bigger than a cupboard, and that the quarter-plates had to be developed in the huge tank intended for large X-ray plates.

Once when the physicist who was helping the doctor was washing a plate, it slipped from his grasp and sank to the bottom. The tank was so deep that even when he groped for it with the full length of his arm it was out of his reach. So he undressed and practically dived for it. He was just drying himself when a nurse hurried in, stopped dead, and nearly dropped the X-ray plates she was carrying.

A few weeks later, de la Warr had a phone call from the doctor who was using the Camera and who now sounded in distress. His own Chief, he said, had been hauled over the coals by superior authority. Orders had been issued that the Camera was to be removed from the hospital at once.

Naturally de la Warr was dismayed, and he asked for further information. None was given him officially, but he gathered that strong undercurrents were at work. The doctor who had been using the Camera had been given an ultimatum to break off all contacts with the de la Warrs or resign his post.

None of the mute evidence in the 400 photographs was investigated. Once again the axe had descended, and de la Warr brought his Camera back home.

CHAPTER THREE

ONCE MORE UNTO THE BREACH

> The inventor tries to meet the demand of a crazy civilisation. Society is never prepared to receive any invention. Every new thing is resisted, and it takes years for the inventor to get people to listen to him and years more before it can be introduced.
>
> THOMAS EDISON

SOME considerable time before King George VI underwent his lung resection operation there was great public anxiety due to the newspapers reporting Buerger's Disease in his leg. Since a person's radiations are transferred to the emulsion when an ordinary photograph is taken, de la Warr tried the experiment of tuning his Camera not with a blood specimen of the King (which he had not got) but with a photograph of His Majesty.

This answered the purpose tolerably well. The diagnostic instrument indicated that he was suffering from Buerger's Disease in the posterior tibial artery of the leg. Photographs were taken with the Mark I Camera on the dial setting for Buerger's Disease in this region, and that shown in Fig. 43 confirmed the presence of the disease in that artery.

To check the correctness of the vibrational rate on the dial setting, another photograph was taken on the same rate. This time the blood specimen of another sufferer from this malady was used, with the Camera still tuned to detect Buerger's Disease in the posterior tibial artery. Fig. 44 shows the result. It indicates advanced thrombosis and clearly shows the obstruction forming in the artery.

Not having a blood specimen of the King, it was impossible to make a really satisfactory comparison, but at least it showed that for the purposes of the Camera an ordinary photograph would act as a link, though not such a good one as a blood spot.

He approached a commercial firm which contacted a young doctor of considerable character and ability on the staff of a big general hospital in London. This doctor consented to meet de la Warr. He examined some of his photographs of disease conditions which had been taken from blood specimens, and overcoming his rather natural prejudices he not only remained polite but even lent a sympathetic ear to the story of how the discoveries had been made—which to any medical man who had never heard of such things must have sounded well-nigh incredible.

Gaining confidence at his first interview, he agreed to visit the laboratories and see the apparatus with his own eyes. De la Warr knew the risk, for his work was largely in the world of atomic physics and most doctors are not very good physicists. But it was a risk which had to be taken.

Some of the doctors he had met were men whose minds were not fully open to completely new ideas, but this one was different. Perhaps the obvious sincerity of the de la Warrs impressed him; possibly the white-coated staff, men who were obviously not cranks, gave him confidence. However it may be, he agreed to take some photographs himself immediately.

Having satisfied himself that the Camera really did work, he began to come down to the Laboratories once a week until he was convinced that the claims made for this new branch of science were just ones; and also that he could operate the Camera himself with the aid of a physicist.

At this point he decided to ask permission from his Chief at the Hospital to import the Camera and keep it there under lock and key. His request was granted.

Some exciting months followed. The de la Warrs made frequent trips to London with their physicist to give what help was needed. Case after case in the hospital was photographed from the patient's blood specimen and the results were checked by X-ray photography, biopsy or autopsy.

More than 400 photographs were taken with excellent results. In a few cases the vibrational rates, which as explained earlier had already been fixed provisionally, had to be modified, but in most instances they were sufficiently accurate, and on the whole things went very well.

Among the first batch of cases chosen from one of the hospital wards was one of tuberculosis of the lungs (Fig. 41), and one of gastric ulcer (Fig. 42).

In Fig. 41 the Camera was tuned to 'Tuberculosis of the Lungs', and the dial setting for this was 402776. In these photographs, which are prints from the plates and therefore 'negatives', the disease radiations from the affected organs appear as large white patches. Notice that this is unlike X-ray photography in which the bones and more solid parts of the anatomy stand out clearly. What is being photographed is distant *radiation*, so that accuracy and clearness in anatomical detail is not to be expected at this early stage of the Camera's development.

Much was learned from these photographs taken in hospital. It became clear, for instance, that three factors entered into the taking and the interpretation of radiation photographs. These are: the vibrational rate which is used on the dial setting, the intensity of the disease radiation, and the anatomical interpretation depending upon the structural features of the radiation pattern.

The third factor is easily the least important; on the other hand, the first factor is vital. As explained earlier, a disease changes its rate of vibration as it develops, and if a series of photographs is taken on slightly different rates, that which gives the clearest picture indicates just what stage the disease has reached.

All the cases chosen for photographic detection in the hospital were subjected to the usual thorough routine tests. The checking up was sometimes done before the Camera was used and sometimes afterwards, so that on some occasions a case was dealt with 'blind'; that is to say, the doctor had no previous certain knowledge of what the Camera should show.

In most cases the results were very satisfactory, but there were some exceptions in the case of bronchial carcinoma. The difficulty here was to establish a vibrational rate which gave consistent results. At length it was realised that there were many stages and ramifications of cancer of the lung and that several rates had to be used before a correct picture could be obtained.

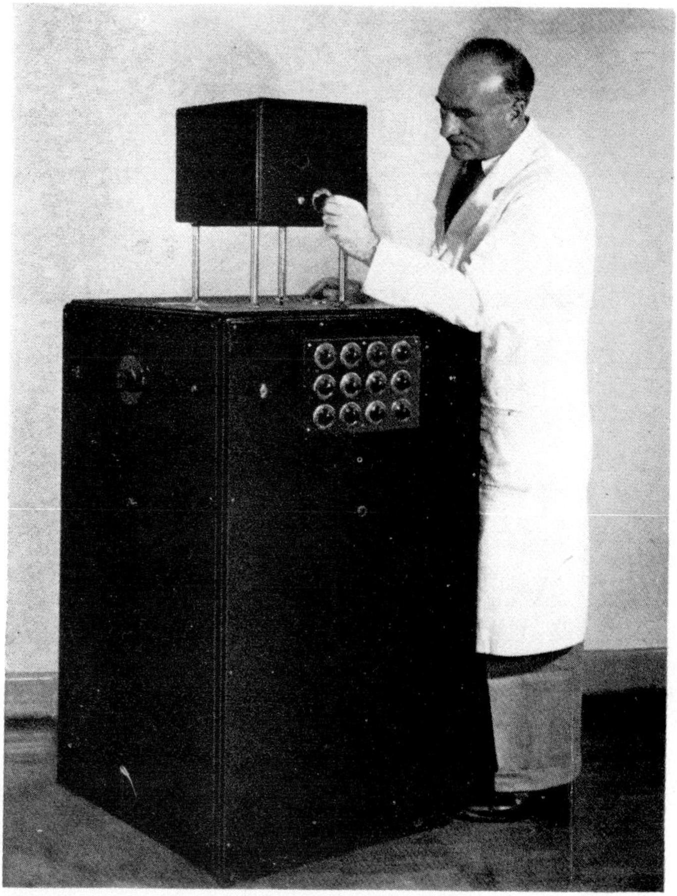

FIG. 40. The Mark I Delawarr Camera
and Mr Stevens, instrument maker

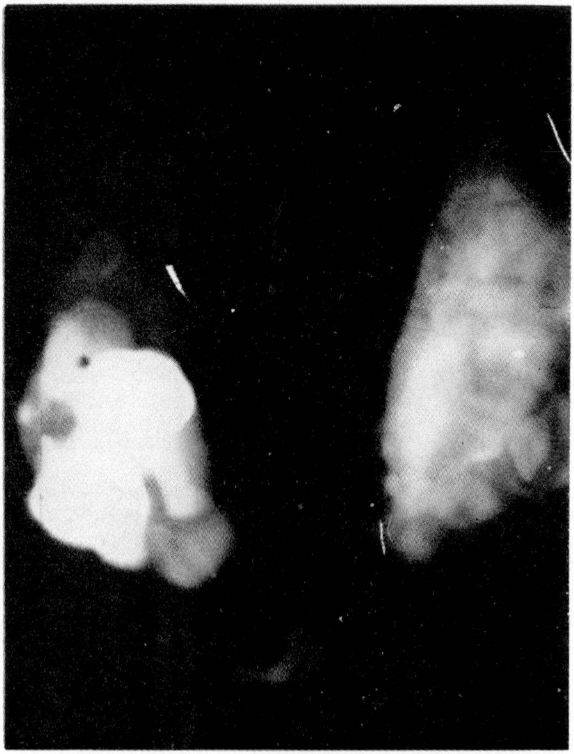

FIG. 41.
Tuberculosis of the Lungs

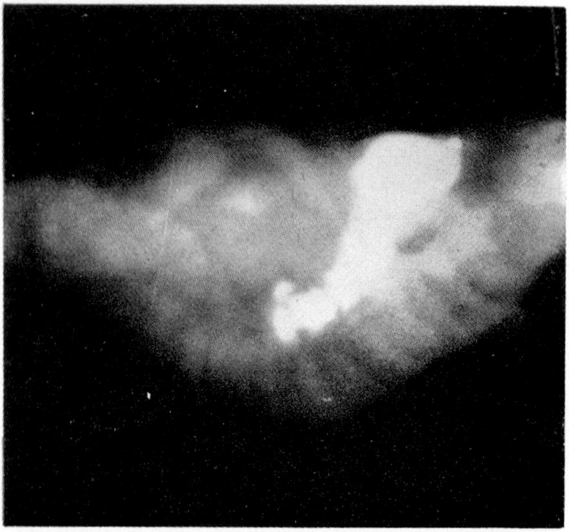

FIG. 42.
Ulceration of the Stomach

Meanwhile de la Warr's debts had been mounting, and unless the Laboratories were to close down it was imperative to get financial help. It was felt that at this stage the London Cancer Hospital would surely be interested, and so a doctor and another man representing the de la Warrs arranged an interview with the doctor in charge of cancer research. Photographs and other evidence were submitted.

The interview was brief. They were told that dozens of similar applications were received every month and that unless the claims were submitted in writing together with full particulars of the apparatus and an explanation of how it worked, nothing could be done.

To this they replied that by a process of selective tuning, cancer in any cell-group of the body could be detected and actually photographed. This could be demonstrated at any time under test conditions, but the inner working of the Camera was complicated and original, and as its principle was not clearly understood it could not be written up satisfactorily from a scientific point of view.

Thereupon the negotiations came to an end. The hospital official made it clear that he did not believe a word of it, and the fact that the principle of the Camera was not understood seemed to convince him that the whole thing was a fraud. Once again de la Warr was thrown back on his own resources.

He realised that if he made a premature disclosure of the Camera's inner construction, imitations would come on the market and he might not even be able to repay the money he had borrowed.

However, at this juncture hope appeared from a new direction. A far-sighted and sympathetic acquaintance in the Secret Service was quick to see the great possibilities of the Camera for use in war. He arranged for the Director of the Royal Naval Research Laboratories to come down to Oxford and talk the matter over.

This man arrived with his chief physicist and the whole thing was discussed in a very friendly way. He agreed that it was unreasonable to expect to be shown the inside of the Camera, but before coming to any decision he asked for his physicist to be allowed to operate the instrument personally for a fortnight to convince himself that it fulfilled the claims made for it. De la Warr consented to this test and a little later the physicist came down.

The situation was a most peculiar one. The physicist found himself very much in the position of a nineteenth-century electrical expert who had been handed a radar set. The principles on which the Camera worked were quite foreign to him. It was impossible to explain to him *why* it worked: he was simply shown *how* to work it.

To de la Warr there was no longer anything strange about its working, but to the physicist it was either pure magic or exceedingly clever trickery. Trained in electro-magnetic principles, he was of course extremely sceptical of de la Warr's theories. So much so that he did not believe there could possibly be any such radiation from a blood specimen, as de la Warr claimed to have discovered.

This was obvious when he tried again and again to take a photograph of a patient's condition with the Camera correctly tuned but without a blood specimen.

To those who watched him he seemed to spend most of his time trying to disprove the evidence which the Camera so obstinately provided when the necessary conditions were fulfilled. Whenever for some reason the result was negative, his face lit up with a glow of hope. But if a photograph appeared on the plate he frowned in perplexity, wondering how he had been deceived. To take some really good photographs seemed to be the last thing he wished to do.

Nevertheless, a few good photographs were taken as well as a great many bad ones. Then a peculiar thing happened. In the course of his interminable juggling with 'blood specimen on', 'blood specimen off', he took a photograph when he should have obtained a blank. This occurred while he was taking a rapid run of exposures.

At the time de la Warr could not account for this result, but later on he discovered that 'residual images' were responsible for it. One had to wait a little while after taking a photograph to allow the energy of the image inside the Camera to fade away.

It is like this. When the energy which activates the Camera is switched on for an exposure, a three-dimensional image is formed inside the Camera. This image is composed of energy particles which remain active even after the apparatus is switched off. Their period of remanence may be as much as three-quarters of a minute, and if another plate is exposed too soon the original image is obtained again.

But to the physicist, here was something which had happened when it should not have happened. He made it clear that he regarded the Camera with deep suspicion, and left the Laboratories.

In spite of this contretemps de la Warr felt that something had been proved which the Naval Intelligence would find well worth investigating. And so he wrote saying that if the D.N.I. would finance him for three months, he would undertake to provide sufficient evidence to enable them to put the matter before their Chief, Sir Henry Tizard.

The reply to this was: 'On the present evidence, I could place no confidence in your apparatus.' It was clear that the physicist had reported unfavourably, and yet another costly experience drew to a close.

At this point in his career de la Warr decided that if necessary he would not only allow some scientist of repute to take the Camera away and use it, but even to look inside it and have a full explanation of how it worked. For by now he was forty-seven years old and unless he could win some official support he might fail to get things moving in his life-time.

Being a Fellow of the Royal Society of Arts, he thought it might be possible for him to read a paper to this Society and under its aegis to get a hearing elsewhere.

In October 1951 he called on the Secretary, showed him some of his photographs and briefly explained the situation. The Secretary was polite but quite incredulous. However, he agreed to come down to Oxford and apply a few tests as a first preliminary.

When he visited the Laboratories the first tests did not go as smoothly as had been hoped because residual images showed themselves again and it was on this occasion that they became evident.

The next test agreed upon was for the Secretary to load four unexposed plates in four opaque plate-holders or cassettes, and in one of them to place a copper sulphate crystal mounted on a card. The de la Warrs had to discover by photography which cassette held the crystal. They were correct on the first run of four, but incorrect on the second run. The photographic image appeared in the wrong cassette, and things looked black indeed until the explanation was found.

At a certain rotational position a crystal becomes a resonator, and when the pile of four loaded cassettes was ready for use with the crystal placed, without the de la Warr's knowledge, in the second cassette from the top, it happened to be in its position of natural resonance. The cassettes were excellent resonant cavities, and the energy from No. 2 was also found in No. 1. Consequently when No. 1 was placed in the Camera it picked up energy particles from No. 2 and an image was recorded on this first cassette.

When this was sorted out and the conditions changed, the test was accepted.

The Secretary then reported the matter to his Chairman and it was agreed that a certain physicist who was on the Council of the Society should be asked to meet de la Warr and talk the matter over. For various reasons this meeting did not take place, but eventually it was decided that de la Warr should be given further tests to be carried out under the Secretary's supervision. This was in November 1951.

The tests were carried out, but they were not entirely satisfactory due to the inevitable differences of opinion about the interpretation of the results. This is a common occurrence when one of the parties is more experienced in the interpretation of photographic images. But in any case, photographs had been taken under test conditions by methods quite unknown to science and were plainly visible. However, after this neither the Society's Chairman nor the physicist would consent to meet de la Warr and one more approach was barred to him.

He then submitted a paper to the Society, but it was refused on the grounds that he had not advanced sufficient proof of his claims. When he insisted that his discoveries were of the greatest practical importance and that the photographs had been taken in precisely the manner which he had described, it was eventually agreed that he should undergo still further tests.

The tests suggested were:

Firstly, using a blood specimen in the usual manner to take five photographs in succession with the Mark I Camera and obtain any photographic reaction whatever.

Secondly, replacing the blood specimen with some red ink to repeat the first test in precisely the same manner.

These two tests were so elementary and so inadequate for proving the possibilities of diagnosis from a blood specimen that de la Warr proposed a third. This was to prove that the photographic reactions obtained in Test 1 were related to the subject to whom the blood specimen belonged by having the subject rotate his body while the exposure was being made. For, as you will remember, if a man turns his body while a photograph is being taken it will blur the image.

Arrangements for these tests and the choice of suitable subjects for it were at length made, but here unhappily de la Warr encountered yet a fresh difficulty. He had naturally told the Royal Society of Arts that Naval Intelligence had carried out some tests on his Camera, and when the Society's physicist requested that the same N.I. physicist who had been so suspicious of the residual images should represent the Society in these later tests, de la Warr had to agree. Small wonder that nothing further happened.

At last he sent a strongly worded protest to the Society asking for a categorical acceptance or refusal to proceed. To the credit of the Society, the Secretary then wrote saying that a committee to whom the negotiations had been reported had agreed to appoint a panel of three scientists to investigate his claims.

Once again de la Warr dared to feel optimistic, but a few days later the Secretary wrote to say that none of the three scientists on the Council was willing to undertake the investigation and that possibly another of the learned societies would be more suitable!

CHAPTER FOUR

CRITICS AND GREMLINS

> It is difficult to conceive anything more scientifically bigoted than to postulate that all possible experience conforms to the same type as that with which we are already familiar, and therefore to demand that explanation use only elements familiar to everyday experience.
>
> PROFESSOR P. W. BRIDGMAN in *Logic of Modern Physics*

IT seemed to de la Warr that the veterinary field was a promising one, especially in regard to foot and mouth disease. Early in 1953 some information was published about the ravages of this malady. Since the beginnings of the outbreak in November 1951, some 50,000 cattle and sheep and 11,000 pigs had been slaughtered by order of the Ministry of Agriculture and about £1,000,000 paid to farmers in compensation.

Germany, where the outbreak originated, reported a loss of £12 millions, France £10 millions, Belgium £7½ millions and Denmark £5 millions. One of the leading experts in foot and mouth disease, Professor Ramon, declared that unless some more efficient means of coping with the scourge could be found, the world's stock of cattle would be faced with 'almost universal disaster'.

There is still controversy about the possibility of inoculating cattle. It seems that the vaccine has serious disadvantages. One is the time-lag after inoculation which allows the disease to spread rapidly. Another is that each individual strain of the virus needs a special vaccine and that the wrong one may be quite useless.

De la Warr felt that his methods should be very well suited to the case, since his Camera could confirm the disease even in its incipient form. In fact it could photograph foot and mouth disease in all its various stages and even detect its presence in the soil, thus rendering it possible to prevent serious outbreaks. Even if the Camera were used only after the beginning of an outbreak, it would provide a means of avoiding the terrible indiscriminate slaughter of merely suspected members of the herd by testing each animal separately.

Nor was there any reason why the new technique of early treatment by Colorscope should not be used in the attack on the disease. Why not develop a system of prophylactic irradiation?

In December 1951 de la Warr wrote to the Secretary of the Royal College of Veterinary Surgeons explaining briefly his work and offering to submit to an investigation, but his offer was not accepted.

In March 1952 an opportunity arose of showing what the Camera could do in the veterinary field when radiation photographs were taken detecting the presence of a piece of iron wire and a stone in the stomach of a cow which was twenty miles away from the Camera. (Fig. 45.) These photographs and others were sent in by a veterinary surgeon to the Royal Veterinary College, with corroborative evidence. Fig. 40 shows the iron wire and the stone after removal by operation.

There was no response from the College, nor was the surgeon or the owner of the cow approached for corroborative evidence.

An approach was then made to Dr Galloway, Director of Research into Foot and Mouth Disease at the station near Pirbright in Surrey. Once again de la Warr offered his services and the use of his Camera, but again without success.

On the other hand there were men in responsible positions who were quick to see the possibilities of his work and who gave him help and encouragement. One of these was Kenneth Walker, the well-known writer, a Harley Street specialist and a former member of the Medical Research Council. He paid several visits to Oxford and soon saw the importance of what de la Warr was doing. To *The Medical World* he sent an article which de la Warr had written on the basic condition of resonance which exists between a patient and his blood specimen. It was rejected, but later it was acknowledged in *Nature*.

In August 1953 another Press conference was held which Kenneth Walker attended with three other doctors who had had practical experience of the work, particularly with the Camera. The M.P. for Newport was there, and Brigadier R. C. Firebrace was in the chair.

Due to the multiplicity of radiations from the assembled company, the Camera could not be shown in action, but it was exhibited and discussed. The Press men with their technical advisers took notes and left by the evening train.

This time three papers published articles. The *Sunday Pictorial* ran a full page under the headline 'Moonshine', *Time and Tide* printed an article captioned 'The Alchemist', and the *Birmingham Post* gave a creditable report. But evidently Fleet Street had decided 'against'.

Clearly, until something appeared in the technical Press it was no use hoping for publicity elsewhere. How was this to be achieved? The experiments on animals were too incomplete for acceptance by *The British Medical Journal* or *The Lancet*. The results of the experiments on basic phenomena, using X-ray plates but no Camera, had been printed and circulated to thirty different universities without eliciting a single enquiry.

The only documented evidence was that relating to the Mark I Camera, and since this was bound up with the whole future of medicine it was decided to approach the Medical Research Council. So detailed drawings, specifications and modus operandi were prepared, and in September 1953 Kenneth Walker sent them in to the Council.

In October the Council wrote saying that if de la Warr believed he had made a discovery of any importance he should publish a full and frank account of it in some technical journal. De la Warr asked for the name of any journal that would be agreeable, but he received no assistance in the matter.

The financial burden had now become so heavy that Lord Nuffield was approached and persuaded to visit the laboratories.

On his first visit it was difficult for him to grasp the extent and importance of the work which was shown to him, but he offered to arrange for a well-known medical man to come down and try to form an opinion on the medical aspect of the discoveries.

Some time later Professor M. arrived with two young physicists, but having no previous knowledge of this new branch of physics they found it very hard to accept de la Warr's claims on the inevitably scant evidence which was presented to them during a stay of two hours.

They suggested a number of tests with blood specimens. For example, if a dozen blood specimens were handed to de la Warr could he determine which of them were from tuberculosis patients?

On the face of it this appeared to be a test in line with his claims; nevertheless, he had to refuse it for two reasons.

The Camera will detect the mechanism of infection at work in the force fields of a patient's body-cells and good positive photographs can be taken of threatened infection. But the alteration in cell structure might not be clinically observable for some months and in the meantime the photographs would be at variance with an ordinary medical diagnosis. Clinical cases of T.B. would of course photograph well, but it is always hard to distinguish between clinical and incipient infection.

The second reason was that unless the Camera team could identify the patient, the apparatus could not be tuned. It was necessary, therefore, to know the name and address of each of the people to whom the blood specimens belonged.

This latter condition seemed unreasonable to the Professor, who was evidently thinking in terms of blood analysis and who could not see what the name and address of the patient had to do with it.

Possibly the reader will feel the same, unless he remembers that the Camera is an artificial aid to the operator's E.S.P. which probes the force field of the patient and which cannot do this until the ordinary mind is told what has to be done and is given sufficient information to initiate this task. In fact, the ordinary mind has to be *briefed*.

It was then suggested that if blood specimens of fifteen guinea-pigs were sent to the Laboratory and then, subsequently, one of them was killed, the Camera might be able to discover which one had died.

This seemed a reasonable test, and it was agreed that a rehearsal of it should be carried out in case of some unforseen difficulty.

A preliminary run was made with four guinea-pigs housed in a different part of Oxford. Then one of them was killed. When photographs were taken with the Mark I Camera to see which of the four blood specimens failed to respond, all four responded equally! And when the experiment was repeated during the next few days, none of them would respond.

Here was another obstacle. Due, possibly to the animals being in metal hutches, or possibly, to the close inbreeding of the guinea-pigs it was apparently impossible to detect the radiations from each animal separately. Whatever the explanation, the test had to be refused. The doctor on de la Warr's staff had to tell the Professor that he would be approached again when a suitable test could be devised.

After some discussion it was agreed that the best way to instil confidence in such people as the Professor would be to loan the Camera out to teams who had been given short periods of training. They would then learn about these problems at first hand, and surely they would not doubt the authenticity of photographs which they themselves had taken?

But again the gremlin which had begun to plague de la Warr played an impish trick. When training a second team it was found that they could not take any photographs at all!

At that time it was not realised to what extent personal factors counted. When using a radar set, the operator cannot affect its functioning and therefore a set can be made for general use. But the research work which followed this failure to take photographs showed that for some teams such as this one the Camera had to be specially adapted to suit them before they could get any result at all.

After several months the Mark VII Camera appeared. This is an instrument which can be adjusted to the personal radiations of the team which uses it.

At this point another influential friend appeared. Lord Glyn's keen and energetic mind quickly realised the importance of de la Warr's work. When he brought it to the attention of the Atomic Research Station at Harwell, a senior chemist on the staff was asked by the Director to visit the Delawarr Laboratories.

On coming down to Oxford, some of the less sensational claims were explained to him and a simple demonstration was given of detecting the radiation from a copper sulphate crystal with the Mark I Camera.

At this critical juncture the gremlin again put in his spoke. Only heavily fogged plates were obtained. It seemed that in some way the presence of the visitor was affecting the delicate balance of forces. Various insulating materials, such as sheets of perspex were tried without success. Only if the visitor were banished to the next room could a good photograph be taken, and naturally this was not the best way to inspire him with confidence!

It was then suggested that if the new Mark VII Camera were adjusted to the chemist's personal radiations he could perhaps take a good photograph himself. The new Camera was fetched, and each of its components was adjusted to suit the chemist as he stood in front of it. A photograph was taken and with extreme anxiety the plate was developed. This time there was no sign of fogging. It showed a perfect image of the radiating lines of the copper sulphate molecule.

The de la Warrs were naturally much relieved. Not for the first time it seemed certain that their work would now be regarded with less suspicion in high quarters. But the gremlin had another trick up his sleeve.

The visiting chemist had himself taken the photograph, but the plate had been loaded into the plate holder by the Laboratory physicist. The Harwell man now tried to load a plate himself as well as taking the photograph, but he found that he could not do it. Evidently his personal radiations were responsible for the failure.

Here was another quite unexpected complication. Hitherto all the plates had been loaded into the dark slide by the Laboratory physicist and no such difficulty had been met with. But now experiments were made with a number of other men and the surprising fact was discovered that out of twelve persons only two could load their own plates, whereas all of them could expose and develop the plates quite satisfactorily.

What was the reason for this? So long as this problem remained unsolved, they were working in the dark and further progress was precarious. All other work was set aside in an attempt to settle it.

By the summer of 1954 it had become clear that operators did not vary in their ability to trigger off the Camera and to impress an image on the emulsion provided the Laboratory physicist loaded the plate. This naturally suggested the possibility of faking the emulsion during the loading operation and the Harwell chemist soon arrived at this simple conclusion, although the operation could be done in the dark-room in ten seconds.

The seriousness of the situation and the ill feeling from all sides was too obvious to ignore. Fortunately from out of the blue came a young physicist from Cambridge who was in some way 'neutrally charged' and was able to handle the plates after the Laboratory physicist had pre-conditioned them, without discharging them.

He could in fact unwrap a new plate in the dark-room and after pressing it on the arms, hands or feet of the Laboratory physicist, take any photograph he wished with his Mark I Camera. This has been repeated so often that it is now standard routine in such research work as still fitfully proceeds.

Indeed, it was a lucky thing for the de la Warrs and for the fate of the whole enterprise that this Cambridge physicist happened to turn up at the Laboratories when he did. It helped to blunt the Harwell attack, but a great deal more research work will have to be done before a satisfactory outcome can be expected. Until others can pre-condition the plates before loading, or until an automatic process has been perfected for doing this, the Camera is in danger of remaining an enigma. In fact, de la Warr has now withdrawn it from the battle and intends to use it only for research purposes until funds permit a fresh attempt on the problem.

It should be remembered that the image formed on the plate is produced not by light waves as in an ordinary camera, but possibly by an electro-static charge drawn partly from the operator and partly from the subject and then projected by the apparatus onto the plate. The charge itself may be a complex aggregate made up of many potentials which serve to form the pattern of the image. The fact that a 'non-loader' can destroy the image by touching the plate before it is placed in the developing dish increases the likelihood that electrostatic charges are involved.

Static electricity is produced by natural causes, and very little is known about it apart from its general behaviour. De la Warr's work will certainly throw new light on the subject, especially when static electricity is considered in conjunction with his discovery of the magnetic spectrum, as revealed by the critical rotational positioning of a bar-magnet in the Earth's magnetic field.

CHAPTER FIVE

THE FORCES OF CONSERVATISM

> They (the scientists) willingly believe that facts that cannot be explained by current theories do not exist.
>
> ALEXIS CARREL in *Man the Unknown*

THE normal method of judging some scientific discovery is to ask such questions as: Is it a logical development of what is already known? Does it conform to the established laws which govern Matter and Energy? Are the experiments which support the claims repeatable, no matter who conducts the experiment?

If the answer to any of these questions is No, then the claims are regarded with the gravest suspicion.

When the discovery is in line with orthodox scientific tenets, when its ancestry can be seen clearly on the genealogical tree, it responds readily enough to these tests and its validity is easily established. But the discoveries which are described in this book fall into a different category. They belong to the world of energy-patterns, resonances and electrostatic potentials, a world half-way between mind and matter with laws of its own. When the exceedingly delicate instruments which function in this half-way world are subjected to tests appropriate to the known world of matter, the results are apt to be disconcerting.

Once the laws of this half-way world are understood and accepted, it will be a different story, but at present de la Warr is rather in the position of a litigant who has to contest a case before a judge and jury who understand only the laws of another country.

If this 'legalistic' factor were the sole aspect of the situation, he would be in a very difficult position indeed, but fortunately there is another side to the picture. The results achieved by his instruments are too substantial to be dismissed by unbiassed minds as mere trickery. Even if the Camera behaves erratically when non-loaders try to load and expose their own plates, there remains the impressive evidence of the photographs which have been taken by others, including such men as the doctor in the general hospital in London and the physicist from Cambridge.

And besides this, there is the testimony of the many practitioners who for years have been using the diagnostic and treatment sets with very considerable success and who frequently send their more difficult cases for photographic verification. It is inconceivable that all these men should be gross-self deceivers.

Some readers may feel surprised that discoveries of such obvious importance should have met with so much indifference on the part of the physicists and the medical profession. No surprise need be felt, nor should we blame these scientists and doctors for their attitude of cautious scepticism towards something which appears to run contrary to immutable concepts.

In a living entity such as a human body, we can see various classes of cells each of which fulfils some special function. And this no doubt is paralleled in the living entity which we call Organic Life. There are many different classes of men who willy-nilly play some special role. For instance, there are the originators of new ideas, new activities, the preservers of what already exists, and those who play a necessary part in the ebb and flow of plastic circumstance.

Those who preserve what exists naturally gravitate to high administrative posts in organised institutions, for it is in the very nature of such bodies to maintain their status quo and resist the influx of ideas which might disturb or disrupt them. These are the men who give stability to a profession or to a learned society and who ensure that what has been gained is not easily lost. But of course this cuts both ways.

If we take the medical profession, we can see that many excellent things such as the Hippocratic Oath have been preserved with the greatest tenacity, and on the other hand that the admission of revolutionary ideas has been resisted with equal tenacity.

William Harvey's discovery of the circulation of the blood nearly cost him his practice. When Ambrose Paré, surgeon to four French kings and to Mary Queen of Scots, suggested tying up the amputated limbs of soldiers with thread instead of immersing them in boiling pitch, his confrères predicted all manner of dire consequences.

Ignaz Semmelweiss who advocated asepsis in maternity cases was hounded out of Vienna by his fellow obstetricians. Jenner who introduced vaccination was assailed by a host of medical writers. Anaesthesia was denounced as highly dangerous and unnatural, the open air treatment of T.B. was considered a frivolous suggestion when put forward by Henry MacCormac, and Freud was considered little better than a charlatan.

In criticising the critics of progressive ideas, most people forget how many useless or dangerous methods and practices they have succeeded in excluding, or what would happen without their restraining influence. Without barriers, the medical profession might soon find itself in a state of chaos. For centuries quack doctoring flourished, and with no brake on its growth it might again assume monstrous proportions.

But the danger is that these forces which safeguard us from quackery and false ideas may become too strong in this age of specialisation and centralisation of power. It is true that a force of resistance favours growth, as the hard crust of earth helps a growing plant, but if it is too strong, new ideas may be stifled. And this is more likely to be the case in highly organised states such as exist in the Western World today.

PART III

CHAPTER ONE

A NEW PRINCIPLE OF UNIFICATION

> When his (Jacob Boehme's) third illumination took place, that which in former visions had appeared to him chaotic and multifarious was now recognised by him as a unity, like a harp of many strings, of which each string is a separate instrument, while the whole is only one harp.

IF scientific discoveries are sufficiently deep, they will give us a different view of Nature and the Universe and throw some light on problems of causation. They will act as pointers to questions which have always puzzled us. Among these we might list: How is order and harmony maintained in Organic life? What is the origin of form or shape? In what way does Nature carry out her building operations? Is there mind behind natural phenomena, and if so how does it act on matter? What is the origin of matter and energy? What is the true meaning of evolution and what is its modus operandi?

It would be absurd to suppose that any laboratory experiments could find complete answers to such profound questions as these. The most they could do is to give us fresh lines of approach which will take us a little nearer to their solution, and perhaps free our minds from false hypotheses and distorted view-points which bar the way to a better understanding of them. Among these last we can safely mention the 'flat earth' point of view which ignores higher levels of organisation and attempts to account for things on one level in terms of factors on a lower level. This leads to the various mechanistic theories and to positivist thinking which explain nothing at all and which may be described as mentally bankrupt.

Can the discoveries described in this book and the inferences drawn from them offer us something better?

Let us begin with the earlier experiments which discovered an interaction between sound waves and subtle radiations such as those emitted by living cells. Sound waves, as we know, are governed by a Law of Harmonic Relationship under which notes of the right frequency combine into chords, and both single notes and chords combine in a time sequence to produce more complex harmonies. There is also the phenomenon of resonance, in which one note or chord vibrates in harmony with components of notes or chords in a higher or lower octave.

Resonance, it has been discovered, also takes place between vibrations of quite a different order, namely these subtle radiations which have been mentioned so often. There seems to be no doubt that they too are governed by the Law of Harmonic Relationship.

How about the radiations of light, radio waves, and so on, in the Electro-magnetic Spectrum, vibrations in the so-called Ether? It seems certain that they too are under this same law and the Do, Re, Mi, Fa, Sol, La, Si, of music are in fact paralleled by the Red, Orange, Yellow, Green, Blue, Indigo, Violet of the light spectrum.

'Talking with physicists, biologists, and other scientists who are unaware of a crisis in musical thought,' says Paul Hindemith,* 'one is always profoundly surprised to how great an extent they operate with concepts analogous to those in musical creation. We have already seen how time and space have their musical equivalents, and there apparently exist similar equivalents to basic laws in the physical sciences. This could lead us to the belief that there is some foundation in the ancient idea of a universe regulated by musical laws—or to be more modest, a universe whose laws of construction and operation are complemented by a spiritual reflection in musical organisms. The time may perhaps return when musical rules will be, as they were in olden times, an essential part of the code of the physical sciences.'

If 'music' is the key to a deeper understanding of Nature and the Universe, it can be said that modern science has hardly got beyond the stage of a small boy playing the piano with one finger. Science occupies itself largely with the vibrations in the Electro-magnetic Spectrum. These are single notes, but Nature it seems is not much interested in single notes; she deals rather

* *A Composer's World,* by Paul Hindemith.

in wave-forms of subtle radiations which combine under the Law of Harmonic Relationship to produce energy-patterns of extreme complexity.

We can imagine a living entity, such as a human body, as a tremendous collection of orchestras playing in harmony, each cell-group representing an orchestra and each cell a musical instrument. When a cell-group is in a state of health, it will play its score correctly, but if debilitated or diseased it will play flat or even in gross disharmony so that the whole 'symphony' is marred. The new technique of diagnosis and treatment consists of tuning-in to such disharmonies and then introducing a force which changes the faulty music and makes it blend in with the rest of the composition.

We can also look at it in a pictorial way. These multitudinous radiations build up a harmonious energy-pattern which is comparable to a picture by a gifted artist. In ill-health, parts of this energy-pattern are distorted and out of balance with the rest of the picture. The new remedial methods alter these disharmonised parts of the pattern, restoring the original wholeness.

Whichever way we look at it, remedial treatment along these lines is an art as well as a science, which agrees with the well-known dictum of Paracelsus.

These radiations which for want of a better notion we can perhaps imagine as vibrations in a sub-ether, must constitute a 'finer' world to which our senses are quite closed. The sounds we hear in Nature, the lights and colours we see on a summer's day, are so tiny a fraction of the harmonies which are continually being broadcast that there are scarcely words to describe the extent of our blindness and deafness. Rudolf von Urban has said that what our special senses reveal to us as compared with what actually exists is about in the proportion of a single drop to all the water in the Pacific Ocean.

There is reason to suppose that such lowly creatures as worms with which we may compare ourselves, are hardly aware of connectedness. At any moment their tiny spotlight of consciousness is focused on what they happen to be touching, with no realisation that it is related to anything else. We are in much the same position. If we could be even a little more aware of the close-woven network of radiations which fills all space we should no more feel that things are truly separate than we should feel disconnected from our own legs.

And with different faculties we should see *how* things were connected. If we were completely flat creatures living on the surface of a table-cloth we should be unable to understand how one corner of the cloth could ever come in contact with another corner of it. We should need one more dimension to perceive the lady of the house folding up the cloth so that the corners came together. In the same way we need another dimension to realise just how a patient can be connected to his blood spot by resonance.

Resonance is not in the three dimensions of space, nor is it in time. It appears to belong to another dimension which introduces a principle of unity or connectedness that is not present in the four-dimensional continuum.

Things resonate because they are in a certain harmonic relationship, not because they are simultaneous or contiguous. Hundreds of miles may lie between them, nevertheless there is this strange connection.

In chemistry we have what is called 'affinity' between certain elements. The well-known characteristic of valency is the power of one element to combine with others whose radiations are in harmonic relationship with its own. Much the same kind of thing is seen in biological processes where one species of cell is attracted by its 'opposite number'. Resonance also accounts for the specific and localised effects of certain drugs and for the highly selective work which is done by hormones, enzymes, and so on.

Higher up in the scale many other enigmas of natural history can be explained by resonance. For example, the migration of the Pacific Golden Plover which includes a direct sea flight of some 2,400 miles to the nearest land, and that of the eels which make their way down the rivers of Europe to a certain region in the Western Atlantic.

In one of the experiments described in this book, de la Warr established resonance with his own wedding which had taken place twenty-two years earlier. Thus it seems that the past can influence the present and the present can influence the future in ways other than those described in history. And for all we know the present can influence the past. What happens today may be conditioned by what will happen tomorrow.

The conclusion we come to is that our apparent separateness is largely an illusion and that as Maeterlinck saw it, we are 'the cells of an immense organism connected with everything that exists by an inextricable network of vibrations, waves and influences'. They extend into the future and stretch back into the past, knitting up the multi-dimensional continuum into a connected whole. And although in a general sense nothing is isolated, yet connectedness is relative. True separateness between A and B is disharmony of their mutual radiations. On the other hand, if there is resonance between A and B, neither time nor space effectually divides them.

CHAPTER TWO

THE GREAT ARTIFICER

> Everything knows something and has had some experience; and everything is a record of its own experience; knowledge and condition are as convertible as force and heat. . . .
> But some things know more than others. Some have had larger experience and made more of it. Their forms and qualities are the expression of this experience in so far as it has affected them.
>
> SAMUEL BUTLER's *Notebooks*

IF everything can be linked in a fifth dimension, then everything exists in a five-dimensional framework. This framework is the Magnetic Spectrum.

Its projection or 'shadow' in our world of three dimensions is a double vortex shaped like that popular toy of the 1910s, the Diabolo. What its shape may be in worlds of four or five dimensions we cannot tell because our faculties are not adapted to such perceptions. Nor can we have more than a faint glimmering of its meaning and importance. Perhaps it may help to say that but for the Magnetic Spectrum, form would not exist, time would not be co-related with space, and the whole Universe would relapse into primeval chaos.

In the face of such profound conceptions we must realise that our ideas about the Spectrum must be crude and naïve in the extreme, yet even so they may lead us a step towards knowledge of Reality.

The simplest of the elements is the hydrogen atom. If we take the planetary analogy, it consists of a single electron revolving round a single proton nucleus. As we should expect, the Fundamental Rays of more complex atoms such as sodium or chlorine are found a little higher up.

Coming now to molecules, which of course are combinations of different kinds of atoms, we climb the Spiral beyond the region of the atoms. Sodium chloride, for instance, which is a chemical combination of sodium and chlorine, is beyond the sodium and the chlorine positions. It appears to be higher up because its atoms have passed through a certain experience, the experience being the union of the sodium and chlorine atoms into a new form of organisation.

Experience is gained by undergoing significant events. Higher up on the Spiral the events which are possible become more varied, and experience widens. A flower can have wider experience than a molecule or a cell, and an animal can have experience still wider.

Such experience modifies the inner make-up of the entity and hence the radiation it emits, or its Fundamental Ray. So since nothing in the Universe passes through precisely the same experience as any other thing, both its inner make-up and its Fundamental Ray are unique—though in the world of atoms this uniqueness may depend upon differences so slight as to be inconceivable.

It may be hard to agree that each molecule and atom, each proton and electron is unique, but perhaps it is easier if we agree to the proposition that in order to be free from all limitation, the Prime Source must create a Universe which actualises all possibilities of existence.

We can see that an electron revolving in an oxygen atom is undergoing different experience from one which is revolving in an atom of uranium; and even that two electrons in two similar atoms are not quite Tweedle-dum and Tweedle-dee because their parent atoms have not passed through precisely the same sequence of events. Each atom, in fact, is a unique little world, and no doubt the slight variations in the movement of the electrons affects the Fundamental Rays.

Thus the elements seem to be far more individual than they might appear, but this is obscured by our limited faculties which are only able to observe them in immense aggregates. If we happened to be as big as a planetary system, we should find it very hard to believe that each human being is unique.

Higher up on the Spiral it is obvious that each event in a person's life modifies the complexity of his Fundamental Ray. Far more than in the case of an atom, his life is unique and his experiences are by no means the same as those of any other person. Each experience modifies his Fundamental Ray, and since each Fundamental Ray is linked to others by resonance, this is at once reflected in its impact on the world outside him. From this it may be possible to approach the idea that what a man *is* is a most important factor. Nor is Being confined to the human race or even to the animate world.

In the Magnetic Spectrum we can also find clues to the meaning of Evolution and the way in which it comes about.

Scientists find it very hard to reconcile a purely materialistic outlook with what can be observed in Nature. This becomes very apparent when they offer theories to account for the growth of a foetus or the origin and development of life on this planet.

To quote Sherrington again*:

> Nerves seem *for* their purpose constructed in view of what *will* be 'wanted' of them. Before ever they function they grow where they *will* be wanted, they make the 'right' connections, those they *should* make. . . . We say, 'It grows into a child.' Grows? Levers laid down in gristle, becoming bone when wanted for the heavier pull of muscles which *will* clothe them. Lungs, solid glands, yet arranged to hollow out at a few minutes' notice when the necessary air shall enter. Limb buds, futile at their appearing and yet deliberately appearing in order to become limbs in readiness for an existence where they will be all-important.
>
> A pseudo-aquatic parasite, voiceless as a fish, yet constructing within itself an instrument of voice against the time when it *will* talk. Organs of skin, ear, nose, tongue, superfluous all of them in the watery dark where formed, yet each unhaltingly preparing to enter the daylit, airy, object-full manifold world which they *will* be wanted to report on. A great excrescence at one end of a nerve-tube, an outrageously outsized brain, of no avail at the moment, but where learning of a world which is *to be* experienced will go forward. All seems to argue prospective knowledge of needs of life which are not yet but are foreknown.

Elsewhere Sir Charles says that such processes are 'all understandable chemistry', yet in the above passage it seems as if

* *Man on His Nature,* by Sir Charles Sherrington.

something were compelling him to set down an array of facts which by no stretch of imagination could be explained by chemistry alone. No biologist has ever written so strikingly of the enormous weight of evidence in favour of intelligence and prescience in Nature.

As regards the origin of life, Professor E. H. Starling, who represents one particular school of thought, after saying that the beginning of life on Earth was probably the formation of some complex substance which had the power to absorb the Sun's rays and use them to form further unstable compounds, goes on*:

> Once given an unstable system such as we have imagined, the great principle laid down by Darwin, viz. survival of the fittest, will suffice to account for the production from it by evolution of the ever-increasing variety of living beings which have appeared in the later history of this globe.

This explanation would hardly have satisfied William Blake who wrote:

> Tyger, tyger, burning bright
> In the forests of the night
> What immortal hand or eye
> Dare frame thy fearful symmetry.

If we think about the superhuman intelligence and artistry which went into the construction of Blake's tiger, its skeletal frame strictly and beautifully comparable to the main girder of a double-armed cantilever bridge, and remember its grace of movement, its intelligence, its 'fearful symmetry' which expresses its feline nature; if we reflect that a tiger is but one member of the mammalian species and think of the bewildering variety of animals, birds, insects, reptiles, fish, trees and flowers, not to mention such things as micro-organisms; if besides this we remember that Nature is a delicately balanced, self-maintaining organism with great powers of adaptation—if we consider all this and then return to Professor Starling's statement, we shall probably be left with a slight feeling of dissatisfaction.

Something more than the instability of chemicals is needed to account for the marvellous order, balance and beauty of Nature.

* *Principles of Human Physiology*, by Professor E. H. Starling.

'The most irrational theory of all', said Plotinus, 'is that elements without intelligence should produce intelligence.' He is talking about what we have called a 'flat earth' theory.

Unstable matter cannot build itself into different forms of organisation which have meaning and purpose any more than rocking stones could build themselves into the cathedral of Notre Dame. But something at the top of the Spiral Stairway can come down and manifest in our world of space and time. This is Creation. And by responding to influences from higher up, passing through integral experience, that which is lower down can ascend the Spiral. This is Evolution.

De la Warr's experiments led him to formulate a new law: 'All forms of energy in the Universe are but manifestations of one basic Cosmic Energy, or Universal Mind.'

This conclusion has been reached by other thinkers, but assuming that the building forces in Nature are intelligent, and even intelligent to a superhuman degree, there remains the problem of their building technique, and particularly how form or shape arises. A clue to this lies in the directional beams emitted by all forms of matter.

The revolution of a single electron round a proton nucleus in a hydrogen atom results in a very simple radiative movement. Strictly speaking, the shape of the electron's orbit is unknown to us, but the revolution is assumed to be rhythmic, and hence the emitted beam (which depends upon it) is also rhythmic.

In a more complex atom we shall get a rather different effect. The uranium atom, for instance, has ninety-two electron orbits and it may well be imagined that the radiation from it is something far too complex to portray accurately. The radiations from molecules which contain Uranium will be more complex still.

As we ascend the Spiral, from atoms to molecules, from molecules to cells, and so on, simple directional beams are replaced by more and more elaborate wave-forms which intersect and interact at a great many points, and at some indeterminate stage in our ascent we shall find distinct energy-patterns emerging.

The process is something like watching an etcher making a great many fine strokes with his pen. At first the strokes look like strokes and nothing else, but gradually they build up into a definite picture. Nature etches with invisible inks which the new process of radiation photography makes visible. Using subtle radiations, she creates multi-dimensional moulds.

In the world of archetypes we can perhaps imagine the creative forces forming moulds in five dimensions, each mould for a whole species. Both time and eternity are involved. From these archetypal moulds would flow countless myriads of individuals each with its four-dimensional mould governing the pattern of its physical shape from birth to death.

Why does an aconite seed grow into an aconite flower? Because, as Fig. 23 shows, Nature has already etched the flower with a pattern related to other patterns. Everything in Creation reproduces its own general shape or form because this is inherent in the energy-pattern associated with its germ-cell. Each one of us has an unique energy-pattern containing in itself all the complexity of the myriads of intersecting radiations which emanate from us, and this pattern determines our physical shape.

We have grown into this shape and not into the shape of a monster because the archetypal mould holds us firmly to the shape of Man. No evidence has ever been produced that a phylum or a species evolves smoothly into another phylum or species. Macro-mutations occur by sudden steps. The most that evolutionary forces can do is to produce micro-mutations within the archetypal limits as if in preparation for the next 'jump'.

This is supported by the nature of the Magnetic Spectrum, the uniqueness of everything which exists and by the hypothesis that Creation involves the working out of all possible experience.

If we consider the Spectrum with its time vector, we can realise that there must be a definite time period during which all possible modifications of an entity's energy-pattern due to differing experiences can come about. Where an atom is concerned, only a very short period is needed. In the case of a whole species it is of course immensely longer, for it comprises the life spans of many generations of creatures, each a little different and contributing its quota of unique experience.

But even this has its limit. If we take a creature like the Dinosaurus we can realise that after the lapse of millions of years all possible modifications of the dinosaurian energy-pattern were worked out. When that happened the archetypal mould would be broken up and the species Dinosaurus disappeared, never to be recreated by the forces of evolution or creation.

Probably in the course of time the same thing will happen to every species, including Man. And in order to actualise further possibilities of experience, Universal Mind will have to create new moulds and manifest in forms at present unknown and perhaps unimaginable.

CHAPTER THREE

FOUR NEW LAWS OF PHYSICS

> We know so little of the real nature of both the electro-magnetic fields and the so-called 'particles' that to claim them as being of the nature of mind is quite plausible. When we go on to consider the biological world with the indubitable evidence of purpose and the necessary inference of mental fields or patterns to which 'matter' must conform, I think we may feel that there are indeed strong indications that the subsistent world is of the nature of mind. . . . We may regard the subsistent world as the ever-present dominant background of a World Mind to which our individual minds are related in a selective manner.
>
> DR RAYNOR JOHNSON in *The Imprisoned Splendour*

IN May 1950 a Scientific and Technical Congress of Radionics and Radiesthesia was held in London at which de la Warr and members of his staff gave lectures based on four new laws of physics which were formulated for the first time. These were:

1. The Fundamental Energy of the Universe, in the form of energy particles, manifests through any energy-pattern which modulates it according to a Law of Harmonics.
2. All forms of matter radiate a combined wave-form of energy which forms a Force Field Body due to the interaction of inherent radiations.
3. The Force Field Body is related to the atomic structure and acts as a complex frame aerial through which Fundamental Energy duly manifests as matter.
4. The modulation wave-form depending on space-time considerations can have wave characteristics.

These laws, which all subsequent research work tended to confirm, explain something about what happens at the point where the creative force is passing into matter and various forms of energy. They are most interesting in their implications, though perhaps a little hard to understand without some further explanation.

They are concerned with Fundamental Energy, or Universal Mind, which being beyond space and time is 'everywhere' and 'always'. It is a limitless ocean of potential energy, imperceptible to us until it is 'conjured up' into our world of sense perceptions. Then it appears as some perceptible form of energy, or as matter. And this is in agreement with Einstein's Generalised Theory of Relativity which says that matter and energy are different aspects of a certain Something which is the source of them both, and moreover that energy can be converted into matter, and *vice versa*.

Form or pattern is the channel through which this Fundamental Energy makes its appearance. The first Law says that it manifests as energy particles through the intermediary of an energy-pattern. The energy-pattern modulates it in accordance with the Law of Harmonic Relationship.

Although it is by no means certain exactly what happens, it seems probable that there is a 'neutral ether' consisting of countless myriads of inconceivably small particles per cubic centimetre having no electric charge and no mass. Universal Mind places electrostatic charges on these particles when there is a suitable magnetic field so that they become charged particles. The complexity of the charge determines the type of charged particle and its behaviour.

This energy having been 'conjured up', the fourth Law says that it can have wave characteristics. Before this happened it was not regulated by the Law of Harmonic Relationship. It was Einstein's Something, a reservoir of undifferentiated matter/energy, being neither waves, nor vibrations, nor quanta, nor anything else to which we could put a name.

Yet without this Something, sustaining and vitalising all, forms of matter both animate and inanimate as an ocean sustains marine life, the Universe would be dead and indeed non-existent, for it would disappear as a picture disappears from a cinema-screen when the light goes out in the projecting-room. But due to its presence, everything is 'alive' and emits radiations.

The second Law makes this clear by saying that all forms of matter radiate a combined wave-form of energy which forms what is called a Force Field Body, and the third Law points out that this Body and the atomic structure of the radiating elements are related.

Returning to the aconite mentioned earlier, we can see that its Force Field Body as shown in the photograph is acting as an extremely complex aerial receiver whereby it receives vital energy. The shape of the aerial determines the kind of vital energy it receives. The aconite aerial pattern is composed of living cells each acting as an aerial and helping to build up the particular shape of the network which characterises *aconite.* Only energy suitable for an aconite is received through this network; and so it is with each of the billions of plants and other living creatures which compose Organic Life.

No doubt resonance with each plant's aerial is conditioned by the Sun and Moon, the rhythm of the seasons, the position of the plant in relation to the Earth's magnetic field, and many other factors.

As to how a living creature grows after fertilisation, the Force Field Body evidently precedes any physical manifestations—something which in the case of human beings would overcome the difficulty of trying to cram vast series of characteristics into the chemical elements which we call genes and chromosomes.

De la Warr believes that the pattern of the change of energy evoked by the Force Field Body (say of an aconite) is subsequently induced into the physical flower by magnetic induction, or by some other process which remains to be discovered. However it may be, static electricity makes its appearance in the growing plant.

At risk of some repetition we must try to make clear what is this radiation which is emitted by all forms of matter. As we have seen, the forms of radiation which figure in the accepted Electro-magnetic Spectrum and the Table of Charged Particles are associated with the movement of electrons, but what we have called 'subtle radiations' are something different. It seems that they are not caused by electrons changing their orbits or flying out of atoms altogether, but their origin is electrostatic charges which under certain conditions in a magnetic field radiate energy particles.

It is these energy-particles which compose the physical radiation from a magnet, the vital vibrations from cells of living creatures, the energies which are evoked and broadcast by de la Warr's treatment sets.

There is a relationship between the inner construction of atoms and their outward expressions in the form of radiations. The third Law expresses this in the words: 'The Force Field Body is related to its atomic structure.' It also says something about the second way in which Fundamental Energy can manifest. This reservoir of potential energy, which like Atlas supports the world on its shoulders, needs an energy-pattern as a channel through which to make its appearance. An energy-pattern, or a Force Field Body, acts as a complex frame aerial, and it is through this that Fundamental Energy manifests as *matter*.

In *The Nature of the Universe* Fred Hoyle says:

> From time to time people ask where the created material comes from. Well, it does not come from anywhere. Material simply appears—it is created. At one time the various atoms composing the material do not exist, and at a later time they do. This may seem a very strange idea, and I agree that it is, but in Science it does not matter how strange an idea may seem so long as it works—that is to say, so long as the idea can be expressed in a precise form and so long as its consequences are found to be in agreement with observations.

Having seen that the power of human thought can modulate an energy-pattern, it is not so difficult to realise the probable role of Universal Mind in the initial process of the creation of matter. At any point in space where there exists a suitably stabilised magnetic field, it can evoke an energy-pattern through which it can itself manifest as matter, and in the course of aeons a new galaxy comes into being. So the dismal doctrine of a dying Universe is replaced by that of a Universe which is perpetually re-created.

CHAPTER FOUR

THE FUTURE

But contemporary science, by the mere fact of having cut itself sharply off from religion and 'mysticism', i.e. by having set up for itself a definite 'taboo', has become an accidental and unreliable instrument of thought. The constant feeling of this 'taboo' compels it to shut its eyes to a whole series of inexplicable and unintelligible phenomena, deprives it of wholeness and unity, and as a result brings it about that we have no science but have sciences.

P. D. OUSPENSKY in *A New Model of the Universe*

SOME critics while admitting the validity of de la Warr's claims try to belittle his discoveries by objecting that his instruments are mere mechanical aids to psychic perceptions. This may be so, but no one criticises the telescope or the microscope because they are mechanical aids to vision. With these and other optical instruments such as the spectroscope we have learned a very great deal about the Universe and the structure of matter, and we have entirely changed our view of this planet's position in the scale of Creation. It is quite possible that in the years to come when these new forms of radiation are better understood we shall think quite differently about such things as the operations of Nature and the inner workings of the human organism.

Science, which reached its point of maximum 'solidification' in the later Victorian period, is already turning towards less grossly materialist concepts; but these new concepts are far from clear, and as with all solidified systems of thought which begin to break up into hard lumps, there is considerable disintegration. Dualistic thought runs through the sciences like a series of Iron Curtains.

For instance, there is great confusion of thought on the problem of Mind and Matter. Professor Adrian, of Cambridge, has probably done more work on the detection of brain impulses than any other man. He has tapped many thousands of nerve centres in the brain and traced their nerve channels to various parts of the human and animal anatomy, recording the impulses on the electroencephalograph. Yet when lecturing on this subject several years ago in Oxford, he said: 'There is no evidence to prove the existence of a mind.'

On the other hand, Sherrington has written*:

> The mind is something with such manifold variety, such fleeting changes, such countless nuances, such wealth of combinations, such heights and depths of mood, such sweeps of passion, such vistas of imagination, that the bald submission of some electrical potentials recognisable in nerve centres as correlates to all these may seem to the special student of mind almost derisory. . . . The mental is not examinable as a form of energy. That in brief is the gap which parts psychiatry and physiology.

These two contrasting views by two most eminent authorities on the subject show clearly the gulf which yawns between men who are still thinking in terms of the old Victorian formulae and those who realise that they do not cover the facts.

De la Warr's discoveries are a step towards bridging this gulf. The human mind which is capable of modulating an energy-pattern has a direct connection with physical forces. Universal Mind is able to manifest through an energy-pattern as matter. The Delawarr Camera connects a purely mental function known as E.S.P. with an image on a photographic plate. And so on. The No Man's Land between psychiatry and physiology narrows down, perhaps eventually to disappear.

This narrowing of the gap is certain to have an effect on medical science which like physics is divided, roughly speaking, into two schools of thought. There are those who believe that all diseases originate in the mind, and others who attribute illnesses to infections, exhaustion of the body, exposure to cold and damp, and other purely physical causes. No doubt both schools of thought are partly right, but a better knowledge of the Force Field Body should bring these differing views into line, and once the new methods of diagnosis and treatment are

* *Man on His Nature*, by Sir Charles Sherrington.

established we should have a medical science which makes less use of crude drugs and relies more on the curative forces of Nature.

The de la Warr instruments provide us, in effect, with a sixth sense which is able to penetrate the crust of gross material results and see what is happening a little higher up on the ladder of causation. With such a faculty it should be possible to develop preventive medicine and also resolve many an ancient controversy.

Are homeopathic dilutions really effective in treating diseases? Have herbs the curative properties which many claim for them? Does X-ray therapy permanently damage the Force Field Body? Is there any virtue in so-called spiritual healing, or the Laying on of Hands? All these things should be capable of scientific verification.

The same thing applies to foods. Apart from discovering by analysis what chemical essentials of diet may be lacking in canned fruits and vegetables, dried and processed foods, etc., it should be possible to test such commodities for vital radiations.

From a nutritional point of view these food preparations are thought to be of equal value to fresh foods, but is this really true? Is there no difference in the vitalising effect between canned peaches and peaches just picked from the tree? Has yesterday's milk lost something which is present in milk warm from the cow? Does the injection of antibiotics into a herd of cows affect the bacterial balance of the animals and detract from the nutritional qualities in the milk yield? Are there stronger radiations in corn grown under natural conditions than in corn grown on land treated with a chemical fertiliser? It may be that we shall soon know the truth about such matters.

We may perhaps venture a few speculations as to where these discoveries may lead in the rather more distant future.

It has often been pointed out that if we possessed eyes which responded to X-rays or Hertzian waves we should see an entirely new Universe. The new radio telescope enables us to 'see' invisible stars, but compared with the subtle radiations with which this book is concerned radio waves, X-rays and Hertzian waves are extremely clumsy media for observation. With a telescope which operates, so to speak, in the sub-ether, who can tell what we shall discover? What shall we 'see', for instance, when we train it on Mars? For one thing we should undoubtedly discover whether or not there is life on Mars similar to our own.

These radiations also stretch back into the past. Thus the skeleton of a megalosaurus or of a pterodactyl is in resonance with the creature which lived millions of years ago, and by analysing the radiations we should be able to discover more about it than can be gleaned from the fossilised bones.

It needs little imagination to realise that vast fields of exploration lie before us and that the knowledge which can be gathered by these new techniques may radically alter the whole pattern of modern scientific thought.

The discoveries briefly outlined in this book show how the material world rests upon a framework of finer vibrations which in their turn have their origin in something still further removed from the material. What lies higher up on the 'Jacob's Ladder' which connects our physical world with the Prime Source we cannot tell, but at least we have a picture of a Universe created and sustained by higher levels of existence.

The dead hand of gross materialistic thought lies too heavily on modern science. Let us hope that the time will soon come when the Message of the Holy Grail is fulfilled:

The restraining and hindering work of the materialist, limited as it is by time and space, must wither and pass away like chaff that is sifted from the wheat.